Biomedical Image Analysis Recipes in MATLAB®

For Life Scientists and Engineers

I dedicate this book to

the dearest memory of my parents Constantino and Carolina,
my beloved wife Brigitte,
my great sons Santiago and Andrés.

Biomedical Image Analysis Recipes in MATLAB®

For Life Scientists and Engineers

Constantino Carlos Reyes-Aldasoro

Lecturer in Biomedical Image Analysis
School of Mathematics, Computer Science and Engineering
City University London
London

Library of Congress Cataloging-in-Publication Data

Reyes-Aldasoro, Constantino Carlos, author.
   Biomedical image analysis recipes in MATLAB® : for life scientists and engineers / Carlos Reyes-Aldasoro.
     p. ; cm.
   Includes bibliographical references and index.
   ISBN 978-1-118-65755-3 (cloth)
   I. American Geophysical Union.   II. Title.
   [DNLM: 1. MATLAB   2. Image Interpretation, Computer-Assisted–methods.
   3. Image Processing, Computer-Assisted.   4. Image Enhancement–methods.   5. Mathematical Computing. W 26.5]
   R857.O6
   610.28′4–dc23
                                         2014049419

Set in 10/11.5pt Meridien by SPi Global, Pondicherry, India
Printed and bound in Malaysia by Vivar Printing Sdn Bhd

1   2015

Contents

Preface, vii

Acknowledgements, ix

About the Companion Website, xi

1 The Basic Ingredients, 1
  1.1 The Matlab Environment, 1
  1.2 Introduction to Matlab, 3
  1.3 Operations with Matrices, 7
  1.4 Combining Matrices, 10
  1.5 Addressing a Matrix, 13
  1.6 Mathematical Functions and Graphical Display, 17
  1.7 Random Numbers, 23
  1.8 Statistics in Matlab, 26
  1.9 Displaying Two-Dimensional Matrices, 29
  1.10 Scripts, Functions and Shortcuts, 37
  1.11 Using Help, 43

2 Introduction to Images, 45
  2.1 An Image as a Matrix, 45
  2.2 Reading Images, 46
  2.3 Displaying Images, 49
  2.4 Colormap, 54
  2.5 Thresholding and Manipulating Values of Images, 59
  2.6 Converting Images into Doubles, 68
  2.7 Save Your Code and Data, 69

3 Introduction to Colour, 71
  3.1 Mixing and Displaying Colours, 71

4 Western Blots, 79
  4.1 Recipe 1: Many Ways to Display a Western Blot, 80
  4.2 Recipe 2: Investigating the Numbers That Make a Western Blot, 93
  4.3 Recipe 3: Image Histograms, 97
  4.4 Recipe 4: Transforming an Image of a Western Blot, 104
  4.5 Recipe 5: Quantification of the Data, 111
  4.6 Recipe 6: Investigating Position of Bands, 121

5 Scratch Wound Assays, 135
  5.1 Analysis of Scratch Wound Assays, 135
  5.2 Recipe 1: Low Pass Filtering Scratch Wound Assays in the Spatial Domain, 139
  5.3 Recipe 2: High Pass Filtering Scratch Wound Assays in the Spatial Domain, 143
  5.4 Recipe 3: Combining Filters and Morphological Operations, 154
  5.5 Recipe 4: Sensitivy to Thresholds and Hysteresis Thresholding, 161
  5.6 Recipe 5: Morphological Operators, 167
  5.7 Recipe 6: Measuring Distances Between Cellular Boundaries, 178

5.8   Recipe 7: Introduction to Fourier Analysis, 187
5.9   Recipe 8: Filtering Scratch Wound Assays in the Fourier Domain, 201
References, 213

6 Bright Field Microscopy, 215
6.1   Recipe 1: Changing the Brightness and Contrast of an Image, 215
6.2   Recipe 2: Shading Correction: Estimation of Shading Component as a
      Plane, 224
6.3   Recipe 3: Estimation of Shading Component with Filters, Morphological
      Operators and Envelopes, 235
6.4   Recipe 4: Mosaicking and Stitching, 247
6.5   Recipe 5: Pixel Intensity and Histograms in Immunohistochemistry, 261
6.6   Recipe 6: Hue-Saturation-Value, 271
6.7   Recipe 7: Multidimensional Histograms, 278
Reference, 289

7 Fluorescence Microscopy, 291
7.1   Recipe 1: Separating and Combining Colour Channels, 294
7.2   Recipe 2: Investigating the Scaling of Values, 298
7.3   Recipe 3: Automatic Threshold Selection, 301
7.4   Recipe 4: Measuring Absolute and Relative Areas, 304
7.5   Recipe 5: Counting Nuclei, 305
7.6   Recipe 6: Quantification of Region Properties Beyond the Area, 308
7.7   Recipe 7: Dividing an Image into Regions, 310
7.8   Recipe 8: Batch Processing and Montages, 316
7.9   Recipe 9: A Myriad of Measurements, 327
References, 341

8 Creating Publication-Quality Figures from Matlab, 343
8.1   Recipe 1: Modifying the Characteristics of the Figures, 344
8.2   Recipe 2: Numerous Plots in One Figure, 352
8.3   Recipe 3: Three-Dimensional Ribbons with Different Annotations, 362
8.4   Recipe 4: Three-Dimensional Graphics, 378
8.5   Recipe 5: Projections, 388
8.6   Recipe 6: Four-Dimensional Data Set Explored, 391

Index, 401

Preface

Most newcomers to the emerging interdisciplinary fields inspired by biological problems, *Systems Biology, Bioinformatics* or *Computational Biology*, will face two difficulties. The first is, of course, that in order to interact properly, a common understanding of the "other" discipline is required. Engineers or computer scientists will need to understand immunohistochemistry or molecular processes at a deep level; deeper perhaps than the statistics and physics that biologists might need to understand. The second problem, which is a bit more subtle, arises at conferences or meetings when two collaborators on the opposite sides of the spectrum engage in a conversation and after some time they begin to wonder if they are speaking the same language. Suddenly, simple concepts do not appear to resonate in the same frequency for the *in-vivo* scientist as for the mathematician. *So when you say 'model' are you really referring to a mouse?* Equally surprised may be the neurosurgeons when they listen to two engineers talking about some *Neural Networks*, which have very little to do with Cajal's discoveries. A table at the end of this section shows a few words, which according to the discipline, can have very distinct meanings.

Within this interdisciplinary atmosphere, the acquisition of images and more generally signals of many dimensions has become a bridge between fields: chemistry, physics and engineering that develop new acquisition technologies, then biology and medicine that apply the technologies to analyse specific problems and then mathematics, computer science and statistics that try to extract meaningful information of the signals. A systematic, algorithmic approach to the storage, processing, classification and statistical analysis, the images can reveal much more than what a human can observe. Furthermore, besides the technological advances that have produced a myriad of imaging techniques, computers have decreased in price and increased in storage capacity and processing power. The need for image analysis from an algorithmic point of view is therefore essential to cope with the rate at which the signals are produced, to provide more reliable and reproducible results and to ease the labour to those who perform manual measurements for hours in front of a microscope or a computer screen.

This book is designed as a tutorial for those researchers who are interested in performing image analysis and whose background is not based on the study of applied mathematics. This book does not assume any previous knowledge of mathematics, physics or programming except probably some basic principles with which most life scientists should already be familiar. The basic concepts of image analysis are presented without delving deep into the theory and applications, and solutions to common biological images are presented. This is the emphasis of the book: to solve common biological problems through the use of image analysis algorithms. In the process of reading this book, a reader should then learn the basics of image analysis and algorithm development so that he or she can then apply these new skills into solving a particular problem in their own laboratories.

The book is based on Matlab, a mathematical language that allows a user to execute flexible and powerful image analysis programmes tailored to the specific needs of the problem. As the nature of Matlab is mathematical, there are numerous textbooks designed for engineers and scientists who may already have a deeper knowledge of mathematics and programming and other concepts such as linear algebra and signals and systems. These books rarely delve into biomedical-specific topics.

This book is designed as a recipe book, in the sense that each chapter will present one self-contained biomedical experiment to be analysed. Chapters 1–3 presents the two basic ingredients of the book: essential concepts of image analysis and Matlab. In Chapters 4–7, algorithm and techniques will be presented as a series of "recipes" or solved examples that show the reader how a particular technique is applied in a specific biomedical experiment. Four biomedical cases are analysed: *Western Blots, Scratch Wound Assays, Bright Field Microscopy and Fluorescent Microscopy*. Each chapter begins with simpler techniques, which then advance in complexity towards the last recipes. Chapter 8 presents some advanced techniques for the generation of publication-quality figures.

The recipes are designed to perform a specific analysis and learn some techniques and algorithms. It is expected that readers will be able to apply these techniques later to their own problems. In addition, they will have the bases to explore Matlab further.

Term	Context	
	Biological	**Nonbiological**
Matrix	The womb; the uterus of a mammal. An amorphous or fibrillar material that surrounds cells; especially the extracellular substance of connective tissue. Also the ground substance in which structural elements (e.g. of a shell, cell wall, etc.) are embedded.	A rectangular array of symbols or mathematical expressions arranged in rows and columns, treated as a single entity, and now usually written within round brackets. Also generally any similar tabulated arrangement of items.
Vector	A bacteriophage that transfers genetic material from one bacterium to another; also, a phage or plasmid used to transfer extraneous DNA into a cell.	A quantity having direction as well as magnitude, often depicted by a line drawn from its original to its final position. An ordered set of two or more numbers (interpretable as the coordinates of a point); a matrix with one row or one column; also, any element of a vector space.
Model	A person, animal or plant that carries a pathogenic agent and acts as a potential source of infection for members of another species. An animal or plant to which another bears a mimetic resemblance.	A simplified or idealised description or conception of a particular system, situation, or process, often in mathematical terms, that is put forward as a basis for theoretical or empirical understanding, or for calculations, predictions, etc.

Acknowledgements

The first ideas for this book were conceived whilst I was a Research Fellow at The University of Sheffield. The multidisciplinary nature of the Tumour Circulation Group lead by Professor Gillian Tozer was a challenging and stimulating environment; it had an ideal research atmosphere where I did my best to learn biology and medicine whilst writing image-processing algorithms. That interdisciplinarity is reflected in this book as many of the recipes here presented follow real problems that were encountered at that time. I am grateful to all the members of the group (Simon Akerman, Meit Björndahl, Jo Bluff, Neil Cross, Rachel Daniel, Phil Davendra, Matt Fisher, Madeleine Fulker, Zahra Gharaei, Olga Greco, Sheila Harris, Jamila Ibrahim, Samira Kazan, Sarah Jane Lunt, Katie Pettyjohn, Lucy Shaw, Abigail Welford, Leigh Williams) but especially to Gillian Tozer and Chryso Kanthou for their help and guidance. In addition, some of the images used in this book were kindly provided by them.

I am also grateful to other friends and colleagues from the School of Medicine: Aliya Ul-Hassan for her help promoting CAIMAN, sharing some test images and regular chats; Peter Scriven and our regular computer-related or western blots exchange; and Professor Stephen Renshaw for the productive neutrophil collaboration. Two members of the Gray Cancer Institute, at that time in London, introduced me to the joys of microscopy. I am grateful to Paul Barber and Borivoj Vojnovic. Gabi Dachs from the University of Otago also kindly donated some images.

My technical skills on image processing were developed during my PhD work at Warwick University; I thank especially Dr Abhir Bhalerao for his guidance and support at that time, as well as for his help with some aspects of this book.

Going further back in time, my interest in image processing started when I worked at Instituto Tecnológico Autónomo de México; I am grateful to Maria Elena Algorri and Federico Kuhlmann.

Finally, I will always be grateful to my family: Brigitte, Santiago and Andrés, for the continuous support during the time I wrote this book, for reading sample chapters to bearing my absence at weekend cinema sessions. My father set the example of writing books in the family; this is my first one following in his footsteps.

About the Companion Website

This book is accompanied by a companion website:

www.wiley.com/go/reyesaldasoro/biomedical

The web site includes the following features:
- Powerpoint presentations of all figures from the book for downloading
- MATLAB® function for downloading
- Example data sets in MATLAB® format

CHAPTER 1

The Basic Ingredients

1.1 The Matlab Environment

The word "Matlab" is an acronym for MATrix LABoratory. Matlab is a powerful technical computing environment and a high-level computer programming language, which provides core mathematics and advanced graphical tools for data analysis, visualisation, as well as algorithm and application development. Matlab is intuitive and easy to use, but it requires the use of command-line interface. That is, contrary to other image-analysis software products, which have interfaces and buttons, in Matlab the user needs to type commands to perform tasks. This may seem daunting for experienced users of graphical/text user interfaces. However, with a little practice the command-line becomes easy to use. Furthermore, graphical user interfaces can be created within Matlab. Many common functions, algorithms, matrix manipulation routines and plotting options are available in the main Matlab program or on one of the many toolboxes that accompany Matlab.

There are several ways to start Matlab: through the "Matlab Icon" on the Desktop, which appears on the Desktop of your computer after installation of Matlab, on Macs it should be under "Applications>Matlab" and on Windows in "Start>Programs". The following figure shows the location of the icons in a Mac Operating System.

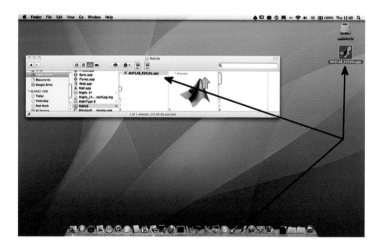

When you start Matlab, one or more windows will appear once the programme has been launched. These may include the 'Desktop', 'Help Window', and 'm-file Editor', as shown in the following figure. You can later configure which windows to launch with Matlab.

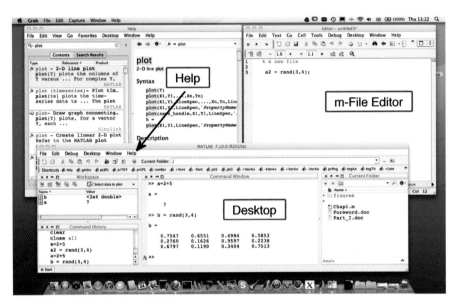

The main window is the 'Desktop', which is subdivided into several sections. The 'Command Window' is where the instructions or commands are written to perform a given task. The user will type the instructions in the 'Command Window'. For example, we can add two numbers, 2 and 5, and store the result in the variable 'a' by typing the following in the 'Command Window':

```
a = 2 + 5

    a =

    7
```

To execute commands, the user must press the "enter" or "return" key after typing each command. It is possible to have one or more commands in a single line, which is sometimes called a line of instructions. After being executed, each line is stored in the 'Command History' window, together with the date when the command was written. This can be very useful for repeating commands that were typed in the past or for finding the commands that generated a certain result. The Command Window is the first place where commands are used, but if a complex task is needed, it may be better to store the commands in a file called an 'm-file', which can be saved, edited and debugged.

The 'm-files' are files of Matlab code that can be used as 'scripts' or 'functions'. Scripts are files in which a series of commands are written. When the file is 'run', all the commands will be executed in the sequence in which they have been written. Functions are different; when a script is executed, any variable that exists already in Matlab may be used. Functions have no knowledge of anything that exists outside the function. In order for a function to know a certain value (stored in a variable), this variable has to be "passed" to the function as input values or 'arguments'. In the same way, any variables that are created inside a script remain in the 'workspace' after the script is executed. With functions the opposite is true; all intermediate variables created inside a function are deleted once the function finishes its execution unless we specifically 'pass' out the variable as an output argument.

To edit the scripts or functions, Matlab provides an 'm-file editor', which is a special text editor that understands Matlab; it highlights the code in different colours so that it is easier to read, it highlights possible errors in the code and suggests improvements and it is very powerful to debug large functions. It even allows the creation of documents. This entire book has been written in the Matlab Editor.

When the commands assign values to variables, a list of variables is displayed on the 'workspace'. Some characteristics of the variables such as their dimensions, value, class ('double', 'single', 'char', 'cell', etc.) are shown. To visualise the details of the variable or to modify some of the contents, a double-click on the variable will display it on the 'Array Editor'. The 'Help Window' provides detailed information about Matlab, functions, editors and toolboxes. New windows can be used, for example when graphics are displayed as will be shown in the next section.

1.2   Introduction to Matlab

In MATLAB everything is a matrix. A matrix should be understood in the mathematical sense as defined by the Oxford English Dictionary: "A rectangular array of symbols or mathematical expressions arranged in rows and columns, treated as a single entity, and now usually written within round brackets" and not as any of the other definitions such as "the uterus of a mammal" or "the amorphous or fibrillar material that surrounds cells". Table 1 in the preface shows some terms with extremely different meanings within the context of the Life Sciences and that of Mathematics and Engineering. Therefore, a single number can be understood as a matrix of dimensions 1×1, i.e. one row and one column. A pair of numbers can be understood as a matrix of dimensions 2×1 or 1×2. This notion of "matrix orientation" (horizontal or vertical) is a very important concept and will be very relevant once operations are performed. Four numbers can be the elements of matrices with dimensions 4×1, 1×4 or 2×2. It is not possible to leave "empty" spaces inside a matrix. For example it would not be possible to place three of the four elements in the first row and the remaining element in the second row. That matrix would automatically become a 2×3 matrix, that is two rows, three columns. The "empty" spaces in the second row would be filled with zeros.

The values of a matrix, that is, the matrix itself, can be assigned to a "variable", which will store the values. The variable can be understood as the name by which the matrix will be known. The basic objective of saving matrices as variables is that they can later be used in conjunction with other variables. For example, to store the value of 5 in a matrix with the name 'a', the following code is typed in the Command Window after the prompt (>>):

```
a=5

    a =

      5
```

After pressing enter, the value is stored, the workspace will show an entry for 'a' and Matlab will "echo" the answer to the command entered in the Command Window. In this case, the echo just repeats what was typed before, but it can be used to find the solution of an equation with more data. For instance:

```
b = 45 + 123 +   a

    b =

      173
```

Notice that we used the matrix 'a' as one of the values to be added and stored in the variable 'b'. In this way, Matlab can be used as a simple calculator. If we are not interested in saving the result, there is no need to assign it to a variable. Matlab will store it in a temporary variable called 'ans' (short of answer).

The order of precedence in which the operations are carried out is the following: exponentiation, multiplication/division and then addition and subtraction. This order can be modified by using round brackets. The order is sometimes referred to with the acronyms "BODMAS" or "BIDMAS", which stand for "Brackets, Order of, Division, Multiplication, Addition, Subtraction" and "Brackets, Indices (powers and roots), Division, Multiplication, Addition, Subtraction". Therefore, the following instruction line:

```
1 + 2 * 3 ^ 4

    ans =

      163
```

is equivalent to:

```
1 + (2 * (3 ^ 4))

    ans =

      163
```

and is different from

```
(((1 + 2) * 3) ^ 4)
```

        ans =

            6561

So far, the command line has been used as a tool to calculate mathematical operations. These calculations can be stored in variables that can be used as part of more complex algorithms; for example, the following line will store the value of the previous line in the variable 'x':

```
x = (((1 + 2) * 3) ^ 4);
```

Notice that this time the line was terminated with a semicolon ";". This final semicolon inhibits the "echo" and thus the final value stored in 'x' will not be displayed. This is useful when variables store large matrices, like the following case:

```
x = ones (15,9)
```

        x =

1	1	1	1	1	1	1	1	1
1	1	1	1	1	1	1	1	1
1	1	1	1	1	1	1	1	1
1	1	1	1	1	1	1	1	1
1	1	1	1	1	1	1	1	1
1	1	1	1	1	1	1	1	1
1	1	1	1	1	1	1	1	1
1	1	1	1	1	1	1	1	1
1	1	1	1	1	1	1	1	1
1	1	1	1	1	1	1	1	1
1	1	1	1	1	1	1	1	1
1	1	1	1	1	1	1	1	1
1	1	1	1	1	1	1	1	1
1	1	1	1	1	1	1	1	1
1	1	1	1	1	1	1	1	1

There are different ways to investigate or look into the values of a matrix. Perhaps the simplest way is to type the name of the variable in the command line, and press enter. Since there is no semicolon at the end of the line, the values of the matrix will be echoed into the Command Window. This method is not very useful for matrices with many rows and columns. For large matrices it is better to use the "Variable Editor", which looks like an Excel spread sheet with the values of the matrix. To open a matrix into the Variable Editor you can use the command 'openvar' followed by the variable of interest like this:

```
openvar x
```

Alternatively it is possible to double-click on the variable on the "workspace". This will open the 'variable editor', which will display the values of the matrix.

To create a matrix with more than one value, you can type its values directly, for instance a sequence of values that start with 1, and increment by one until 9:

```
x = [ 1 2 3 4 5 6 7 8 9 ];
```

Of course, it would be very slow to type each value when you have large matrices. Matlab provides several techniques to fill in matrices more efficiently; for instance, the previous matrix could be created with the following instructions:

```
x = 1:9;
```

or

```
x = linspace(1,9,9);
```

The first command uses the "colon operator", which is a shortcut to indicate a series of values by indicating only the initial and final values; an increment of one is implicit. It is possible to define the initial and final value and increment (lower limit: increment: upper limit) using the colon operator in the following way:

```
z = 0 : 0.1 :20;
```

The colon operator is widely used in Matlab, not only to create matrices, but also to manipulate them. The section "Addressing a Matrix" will explain in more detail the use of the colon operator.

The second command uses a pre-defined Matlab function called 'linspace', which stands for "linear space". As was mentioned previously, Matlab functions "receive" one or more input arguments, in this case the parameters with which the new matrix will be created. These arguments are then used to return a specific output argument, in this case the matrix itself. The command 'linspace' creates a series of values that are equally spaced between an initial and a final value and the number of intermediate points specified by the third argument. This function is useful when the increment between values is not directly known, for instance:

```
x = linspace(1,10,9)

x =

  Columns 1 through 7

   1.0000   2.1250   3.2500   4.3750   5.5000   6.6250   7.7500

  Columns 8 through 9

   8.8750   10.0000
```

This command has created a matrix with 9 values between 1 and 10, the increment between values is calculated by Matlab.

When the spacing is equal between the values it corresponds to a "linear" spacing, but it is possible also to have unequal spacing "logarithmic" spacing with the function 'logspace', for example:

```
x = logspace(0,2,5)
```

```
        x =

    1.0000    3.1623    10.0000    31.6228    100.0000
```

Notice that the space is formed with base 10 logarithms; five intervals between 10^0 and 10^2.

As was mentioned previously, one important concept in Matlab is the orientation of the matrices, therefore

```
y = [1 2 3 4 5 6 7 8 9 10];
```

is different from

```
y = [1;2;3;4;5;6;7;8;9;10];
```

The conversion from a vertical matrix to a horizontal 1 or vice-versa is call "transposition" and is obtained with a single quote after the matrix. For example, when a matrix is created in between square brackets, it is horizontal by default; however, with the single quote it is transposed into a vertical matrix and stored in the variable 'y':

```
y=[1:10]';
```

1.3   Operations with Matrices

It is very important to be aware of the size and orientation of the matrices, especially when they are used as part of an operation. Matrices can be added, subtracted, multiplied or divided in similar ways to numbers. They can also be combined or concatenated to form larger matrices or split to form smaller matrices, and it is possible to "address" only specific values of the matrix, not necessarily contiguous to each other. To illustrate some operations, we will create two matrices and perform a series of mathematical operations with them.

```
x                      = 1:10;
y                      = 11:20;

z1                     =    x +  y;
z2                     =    x -  y;
z3                     = -  x +  y;
z4                     =    x .* y;
z5                     =    x ./ y;
z6                     =    x .^ y;
z7                     = [x     y];
z8                     = [x   ; y];
```

```
z9                          =    [x'    y'];
z10                         =    [x' ; y'];
z11                         =     x .* y';
z12                         =     x'.* y;
```

We will now describe the previous examples:

The matrix 'z1' is the result of the point-to-point (or element-wise) addition of two matrices. Element-wise means that the first element of 'x' is added to the first element of 'y', then the second elements are added and so forth. For this operation to work, it is required that the dimensions of the matrices "agree", that is, both matrices must have the same dimensions of rows and columns. Multiplications are slightly more complicated as we will see later. Element-wise subtraction is performed to obtain 'z2' and 'z3'.

Element-wise multiplication and division are used to calculate 'z4' and 'z5'. Notice that before the operator (the asterisk for multiplication and slash for division) there is a dot ("."). This is very important. When a dot is used before an operator it indicates that the operation should be performed element-wise, i.e. the first element of 'x' is multiplied to the first element of 'y' and so forth. Without a dot, the multiplication operator * would perform a "matrix multiplication", which follows specific rules from linear algebra. To multiply the elements of two matrices in a point-to-point basis you need the dot. In the calculation of 'z6', each element of 'x' is raised to the power of the elements in 'y', i.e. 1^10, 2^11, 3^12, etc.

The examples 'z7' to 'z10' show how the two matrices can be combined. The matrix 'z7' is the concatenation in a single row of the values of 'x' and 'y', one after the other, while for 'z8' the combination will place the values of each matrix in separate rows, one below the other. Notice that the final dimensions of 'z7' and 'z8' are different.

```
size(z7)
size(z8)


        ans =


            1      20


        ans =


            2      10
```

To display all the variables that have been created we can use two very useful commands; the first is 'who', which lists the current variables:

```
who


        Your variables are:


        a                       y       z12     z6
        ans                     z       z2      z7
```

b	z1	z3	z8
	z10	z4	z9
x	z11	z5	

The second command 'whos', which lists the variables, one per line, with basic information of each. It displays the name, the dimensions, the number of bytes and the data type, and in some cases it also displays the 'attributes' of the variables

whos

Name	Size	Bytes	Class	Attributes
a	1×1	8	double	
ans	1×2	16	double	
b	1×1	8	double	
x	1×10	80	double	
y	1×10	80	double	
z	1×201	1608	double	
z1	1×10	80	double	
z10	20×1	160	double	
z11	1×1	8	double	
z12	10×10	800	double	
z2	1×10	80	double	
z3	1×10	80	double	
z4	1×10	80	double	
z5	1×10	80	double	
z6	1×10	80	double	
z7	1×20	160	double	
z8	2×10	160	double	
z9	10×2	160	double	

For the calculation of 'z9' and 'z10' the matrices are transposed before they are combined, therefore they are stored as columns instead of rows. You can verify that the elements of 'z7' are equal to those of 'z10' but with different orientations, and the same applies to 'z8' and 'z9'.

The operations of 'z11' and 'z12' follow matrix algebra; notice that one of the matrices is transposed before the multiplication. The first case is known as the "inner product", and it is equivalent to the sum of the elements of the point-to-point multiplication of the values of 'x' and 'y' (also known as element-wise product). The second case is known as the "outer product" and creates and array of all the possible multiplications between the elements of 'x' and 'y' stored in a 10 × 10 matrix. The products x*x and y*y are not valid because the matrix dimensions "do not agree".

Variable names can have more than one character. In fact, it is better to use sensible names for the variables so that the coding of large algorithms is easier to read; for instance, 'substance_concentration' or 'initial_velocity' are more descriptive than 'x' and 'y'.

1.4 Combining Matrices

As you have noticed by now, matrices are not restricted to having only one column or one row. Matrices can have many dimensions. Indeed, an image is a special case of a matrix, as we will see later. You can create a matrix with several rows and columns by typing them directly; each row is started with a semicolon:

```
matrix_1          = [1 2 3 4;2 3 4 5;3 4 5 6;4 5 6 7; 9 9 9 9];
```

You can create a matrix by using other matrices as with the previous examples (z7 or z8), but you need to be careful with the dimensions of the elements that will form a new matrix. As an example, we can form two new matrices based on 'matrix_1', first by concatenating them horizontally and then vertically like this:

```
matrix_2          = [matrix_1 matrix_1]

matrix_3          = [matrix_1;matrix_1]
```

matrix_2 =

1	2	3	4	1	2	3	4
2	3	4	5	2	3	4	5
3	4	5	6	3	4	5	6
4	5	6	7	4	5	6	7
9	9	9	9	9	9	9	9

matrix_3 =

1	2	3	4
2	3	4	5
3	4	5	6
4	5	6	7
9	9	9	9
1	2	3	4
2	3	4	5
3	4	5	6
4	5	6	7
9	9	9	9

In these examples 'matrix_2' has 5 rows and 8 columns, and 'matrix_2' has 10 rows and 4 columns. In traditional mathematical books, there would be large square brackets around the numbers of the matrices to show how they are delimited. These brackets do not appear when the values are echoed into the Matlab Command Window.

Unlike the previous two examples, the following concatenation of matrices will create an error as it is not possible to combine matrices whose dimensions do not agree:

```
try
    matrix_4              = [matrix_2 matrix_3];
catch errorDescription
    disp('Error using ==> horzcat');
    disp('Dimensions of matrices being concatenated are');
    disp('not consistent.');
end
```

```
          Error using ==> horzcat
          Dimensions of matrices being concatenated are
          not consistent.
```

In the previous lines we have used an advance feature of programming languages called "exception handling" or "error catching", which is designed to deal with errors. Matlab handles exceptions in the following way: all the lines of code that are placed inside a block delimited by the statements 'try' and 'catch' (only one line in this case) are under a special consideration. If errors occur inside this block, Matlab will not return an error message or crash a process if it was run within a function or script. Instead, Matlab will "catch" the exception and record the error in the variable placed immediately after the command 'catch' ('errorDescription'), and will execute instead the statements in the block delimited by 'catch' and 'end'. In this case we only displayed an error message, but in other cases it is possible to indicate alternative commands.

The error messages in Matlab in some cases may seem a bit cryptic, but it is important to read them to understand the possible sources of any errors. In this example, the inconsistent dimensions of the two matrices to be combined created the error. We can investigate the variable 'errorDescription' of our previous example by typing the name of the variable on the Command Window like this:

```
errorDescription
```

```
          errorDescription =

          MException with properties:

                 identifier: 'MATLAB:catenate:dimensionMismatch'
                    message: [1×61 char]
                      cause: {0×1 cell}
                      stack: [5×1 struct]
```

We can observe that 'errorDescription' is a variable of the type 'MException', which is a special arrangement of fields for error recording. These arrangements of many variables under a single name are called 'structures' or 'structs' and are common to many programming languages.

The field 'message' is of particular interest; we can investigate its contents typing the name of the variable, then a dot and then the field of interest like this:

```
errorDescription.message
```

```
     ans =

     Dimensions of matrices being concatenated are not
         consistent.
```

The message describes the cause of the error, which, as we already knew was the different sizes of the matrices.

You can also combine matrices and apply mathematical operations at the same time. For example, we can take one matrix, apply different mathematical operations to it and store all results in a new matrix. Let's illustrate this example by modifying 'matrix_1' with several operations like this:

```
matrix_4            = [matrix_1  matrix_1+20 matrix_1*3   ;...
                       matrix_1./2  sqrt(matrix_1) matrix_1.^2 ]


  matrix_4 =

  Columns 1 through 7

     1.0000   2.0000   3.0000   4.0000   21.0000   22.0000   23.0000
     2.0000   3.0000   4.0000   5.0000   22.0000   23.0000   24.0000
     3.0000   4.0000   5.0000   6.0000   23.0000   24.0000   25.0000
     4.0000   5.0000   6.0000   7.0000   24.0000   25.0000   26.0000
     9.0000   9.0000   9.0000   9.0000   29.0000   29.0000   29.0000
     0.5000   1.0000   1.5000   2.0000    1.0000    1.4142    1.7321
     1.0000   1.5000   2.0000   2.5000    1.4142    1.7321    2.0000
     1.5000   2.0000   2.5000   3.0000    1.7321    2.0000    2.2361
     2.0000   2.5000   3.0000   3.5000    2.0000    2.2361    2.4495
     4.5000   4.5000   4.5000   4.5000    3.0000    3.0000    3.0000

  Columns 8 through 12

    24.0000    3.0000    6.0000    9.0000   12.0000
    25.0000    6.0000    9.0000   12.0000   15.0000
    26.0000    9.0000   12.0000   15.0000   18.0000
    27.0000   12.0000   15.0000   18.0000   21.0000
    29.0000   27.0000   27.0000   27.0000   27.0000
     2.0000    1.0000    4.0000    9.0000   16.0000
     2.2361    4.0000    9.0000   16.0000   25.0000
     2.4495    9.0000   16.0000   25.0000   36.0000
     2.6458   16.0000   25.0000   36.0000   49.0000
     3.0000   81.0000   81.0000   81.0000   81.0000
```

There are several important things to notice from the previous instruction. First, notice the way the line was written; part of the matrix was written in one line and part in a second line. In order to do this, we indicate that we are going to break a line by using three consecutive dots ('...'). Matlab will then interpret those two lines as a single line.

Second, notice that the numbers have been displayed in a format different from the previous examples where only 'matrix_1' was displayed. Matlab has different ways of storing numbers; if they do not have a decimal value, they are 'integers';

when they can be either true or false, yes or no, 1 or 0 they are called 'logical' or 'Boolean'; when number have decimal places they are called 'double', alphanumeric characters such as letters are called 'char'. These are just a few of the many possible cases of what is called "data types" or "data classes". Matlab will convert between data types automatically in most cases, as in the previous example, so there is no need to worry about declaring types or worrying about which case to use. As a general rule, numbers should be 'double'.

Finally, notice that the squaring of the values used the "." dot before the square operator to indicate that the squaring should be performed to the individual values of the matrix and not squaring a matrix in the way that this can be done in linear algebra. For the multiplication of a matrix and a number it is not necessary to use a dot. The multiplication indicates that each element of the matrix should be multiplied by a factor of 3.

1.5 Addressing a Matrix

In many cases, especially when you are dealing with large matrices as the ones that will correspond to an image, it is important to be able to obtain or modify only a subset of the values of the matrices. Imagine for instance, (a) that you want to zoom to a certain region of an image, (b) you want to subsample your image by discarding odd rows and keeping the even ones, or (c) you may want to select only those experiments for which a certain concentration of cells is larger than a predefined value. All these examples are very easy to do in Matlab by "addressing" a matrix.

The simplest cases of addressing a matrix are to obtain values that are in special positions of the matrix. For example, to obtain a value or range of values of the previously defined matrices you need to type the values between brackets after the name of the matrix with the following format (row(s), column(s)).

```
matrix_1(2,2)
matrix_2(2:3,1:3)
matrix_3(1:2:end,:)
```

```
        ans =

            3


        ans =

            2       3       4
            3       4       5


        ans =

            1       2       3       4
            3       4       5       6
            9       9       9       9
            2       3       4       5
            4       5       6       7
```

The previous examples show three ways of addressing a matrix. First, a single element is retrieved; the value is stored in the second row and second column. The second example retrieves the values stored in the second to third row and first to third column. The third example uses the colon operator described previously to address all the odd rows. The shorthand notation '1:2:end' can be translated into words as: "rows that start at 1, increment by 2 until rows finish". The columns also use a shorthand: the colon operator on its own ":" is equivalent to writing '1:end' or "take all columns of the matrix_3".

In the same way that the addressing of the matrix can be used to retrieve values, it can also be used to change the values of a matrix. For instance, if we would like to change to 0 the first column of a matrix we could type:

```
matrix_1(:,1)              = 0
```

```
matrix_1 =

    0    2    3    4
    0    3    4    5
    0    4    5    6
    0    5    6    7
    0    9    9    9
```

or change all the values of the first row to -1:

```
matrix_2(1,:)              = -1
```

```
matrix_2 =

   -1   -1   -1   -1   -1   -1   -1   -1
    2    3    4    5    2    3    4    5
    3    4    5    6    3    4    5    6
    4    5    6    7    4    5    6    7
    9    9    9    9    9    9    9    9
```

To change a specific region, for example, to leave the border elements unchanged and set all the inside elements of the matrix to 0 we could type:

```
matrix_3(2:end-1,2:end-1)  = 0
```

```
matrix_3 =

    1    2    3    4
    2    0    0    5
    3    0    0    6
    4    0    0    7
    9    0    0    9
```

```
1        0        0        4
2        0        0        5
3        0        0        6
4        0        0        7
9        9        9        9
```

or the opposite, to set the boundary elements to 0 and leave the central values unchanged:

```
matrix_4([1 end],:)      = 0;
matrix_4(:,[1 end])      = 0
```

```
matrix_4 =

Columns 1 through 7

0         0         0         0         0         0         0
0    3.0000    4.0000    5.0000   22.0000   23.0000   24.0000
0    4.0000    5.0000    6.0000   23.0000   24.0000   25.0000
0    5.0000    6.0000    7.0000   24.0000   25.0000   26.0000
0    9.0000    9.0000    9.0000   29.0000   29.0000   29.0000
0    1.0000    1.5000    2.0000    1.0000    1.4142    1.7321
0    1.5000    2.0000    2.5000    1.4142    1.7321    2.0000
0    2.0000    2.5000    3.0000    1.7321    2.0000    2.2361
0    2.5000    3.0000    3.5000    2.0000    2.2361    2.4495
0         0         0         0         0         0         0

Columns 8 through 12

      0         0         0         0        0
25.0000    6.0000    9.0000   12.0000        0
26.0000    9.0000   12.0000   15.0000        0
27.0000   12.0000   15.0000   18.0000        0
29.0000   27.0000   27.0000   27.0000        0
 2.0000    1.0000    4.0000    9.0000        0
 2.2361    4.0000    9.0000   16.0000        0
 2.4495    9.0000   16.0000   25.0000        0
 2.6458   16.0000   25.0000   36.0000        0
      0         0         0         0        0
```

Notice that it is possible to use the 'end' operator as a value. This is very useful when you do not know the dimensions of the matrix, for instance when you are going to process images of different sizes.

There are two more important issues to know regarding the addressing of a matrix. First, it is possible to copy parts of a matrix by addressing the matrix to be copied, for instance:

```
matrix_5                  = matrix_1(1:end,2:3);
```

This function will be handy when dealing with images, for instance in a colour image, and we are only interested in the blue channel, or when we want to zoom into a special section of the image.

In the previous examples we have enlarged matrices by combining them. In this way, we have added elements, appended them to non-existent places of the matrix. We can create matrices of any dimensions by assigning just one value to a position. For example, if we would like to create a matrix with 500 rows, 200 columns and 3 levels (a three-dimensional [3D] matrix, such as an image with three colour channels) we only need to assign a value, usually a 0) to the last position of the matrix:

```
matrix_6(500,200,3)      = 0;
```

We can extend already existing matrices in the same way:

```
matrix_1(10,15,4)        = 0;
```

Second, to remove elements of a matrix, we need to declare those elements as 'empty'. An empty matrix is declared in Matlab as two square brackets and nothing between them ("[ ]"). Thus to remove the first column of a matrix, we could type:

```
matrix_2 (:,1)           = []
```

```
matrix_2 =

  -1    -1    -1    -1    -1    -1    -1
   3     4     5     2     3     4     5
   4     5     6     3     4     5     6
   5     6     7     4     5     6     7
   9     9     9     9     9     9     9
```

To remove the last row of the same matrix we would type the following code into the Command Window:

```
matrix_2 (end,:)         = []
```

```
matrix_2 =

  -1    -1    -1    -1    -1    -1    -1
   3     4     5     2     3     4     5
   4     5     6     3     4     5     6
   5     6     7     4     5     6     7
```

It should be apparent by now that the size of the matrices can be important in many cases, either to know the dimensions so that the matrix can be addressed, or to prepare other matrices that will be used in future commands. To obtain the dimensions of a matrix we use the command 'size' followed by the name of the variable we want to measure the size. It is possible to echo the answer to the Command Windows, but we can also store it in new variables in case we need that value later on. As it was said before, it is convenient to use sensible names for our variables, for example:

```
[numberRows,numberColumns] = size (matrix_2);
```

If you are only interested in one dimension of a matrix, you can specify that dimension by writing down the number of the dimension after the name of the matrix like this:

```
[numberColumns]        = size (matrix_2,2);
[numberLevels]         = size (matrix_2,3);
```

1.6   Mathematical Functions and Graphical Display

In this section we will introduce the basics of plotting, figures and images, and at the same time we will introduce the concept of functions and how they are used in Matlab. Later on, you will learn to write down your own functions and scripts. These functions can also be used to create figures. In Chapter 8 we will show you how to produce high-quality figures suitable for journal submissions.

One of the biggest advantages of using Matlab is the possibility of using mathematical functions, such as logarithms, square roots or absolute values, within matrices. In traditional programming languages, if you were interested in obtaining the logarithm of all the values of a matrix, you would need to loop over all the elements and obtain the logarithm of each element. In Matlab this is not necessary because the functions are applied to the whole matrix in one single instruction.

The following examples will use some of the most common Matlab commands and mathematical functions to illustrate how we can generate and manipulate data with a few simple commands. First, a range of values is assigned to a variable. For example, we can consider these values as the times of an experiment that takes samples of a certain measurement every half a minute (0.5) for a space of 1 hour (0:60). Our variable 'time_range' would be defined like this:

```
time_range             = 0:0.5:60;
```

Now imagine that the observation would behave in a sinusoidal way and is related to the time. We can then use the command 'sin' that calculates the sine values of each of the elements of 'time_range'.

```
observation_1          = sin  (0.5*time_range);
```

Notice that we can have intermediate operations like multiplying by 0.5 before calculating the sine function. To obtain a second observation with the same dimensions as the previous one, but this one formed of randomly distributed numbers, we could use the function 'size' and the function 'randn' in one single command:

```
observation_2          = randn(size(observation_1));
```

The command 'randn' generates random numbers that are distributed following a Normal or Gaussian distribution (mean = 0, standard deviation = 1). The output of the command 'size' is used by 'randn' to generate a matrix of the same dimensions as 'observation_1' with the random numbers. If we were now interested in the absolute value of the observations we would apply the command 'abs' directly to 'observation_1' and store the output in a new variable:

```
absolute_observation_1  = abs (observation_1);
```

The function 'sign' obtains the sign of each element of the matrix, that is, it tells if a value is positive, negative or 0.

```
sign_observation_1        = sign (observation_1);
```

At this point we can realise that observing the actual values of the matrices is not always the best option to analyse our data. Matlab is particularly good to display matrices in many different graphical ways. Some programmers sometimes prefer to use traditional numeric programming languages like "Fortran" or "C", in part because they have complete software solutions that have been developed over many years and they do not wish to migrate to Matlab, and also because in some cases those programmes may run faster than Matlab. However, they prefer to use Matlab to display the output of their programmes, as Matlab is very good for graphical display, and far easier than those programming languages.

The first graphical display that we will show is the one generated with the command 'plot'. The command 'plot' displays the values of a one-dimensional matrix as a graph in two dimensions. One dimension corresponds to the position of the values of the matrix, and the other dimension corresponds to the actual values. Let's observe the plot of one of the matrices previously generated:

```
plot(observation_1);
```

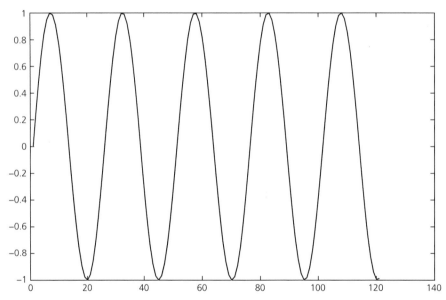

The graph shows a sine curve that starts in 0, oscillates between $[-1, +1]$ and spans 121 time points; the ones that were created by defining the points between 0 and 60. The horizontal axis is defined by the location of the values of the matrix from 1 to 121. However, we know that these observations were done between 0 and 60, therefore we would like our plot to show these values in that range. To combine the time and the values of the observations in one single plot, we need to indicate this by passing two variables as arguments to the command 'plot'; first the variable that will correspond to the horizontal axis and second the one that corresponds to the vertical axis, e.g.

```
plot(time_range,observation_1);
```

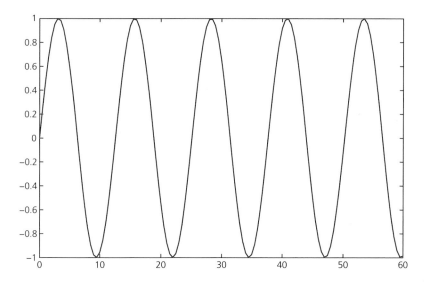

Notice that the graphic is very similar but now the values of the horizontal axis go from 0 to 60 and there is no empty space on the right hand side of the graph. By default Matlab draws the plot with a continuous blue line between points. However, we can easily modify this by adding a third argument to the command 'plot'. For example, we can draw the line in red, with round markers to each point, like this:

```
plot(time_range,observation_1,'r-o');
```

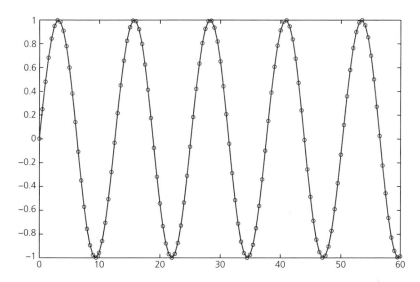

The third argument 'r-o' is the shorthand notation to specify a red colour ('r'), a continuous line ('-') and round markers ('o'). There is a wide range of possible colours, lines and markers to be used. These are especially useful when plotting several lines in the same figure, for example:

```
plot(time_range,observation_1,'r:o',time_range,...
    absolute_observation_1,'b-x');
```

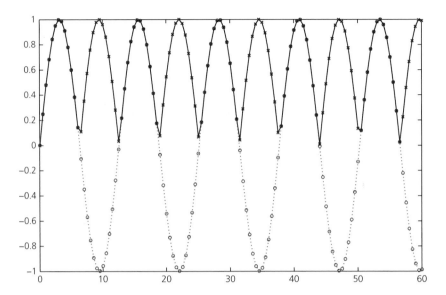

or

```
plot(time_range,observation_1,'r:o',time_range,...
    sign_observation_1,'b-');
```

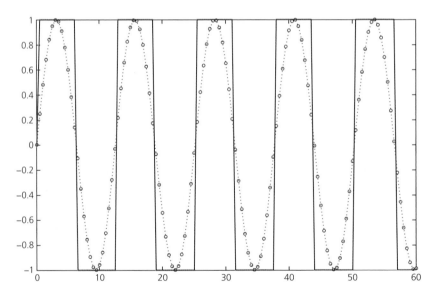

It is possible to combine plots by applying functions or operations directly on the plotting lines:

```
plot(time_range,observation_1,'r:o',...
    time_range,absolute_observation_1,'b-x',...
    time_range,1+absolute_observation_1,'k--d');
```

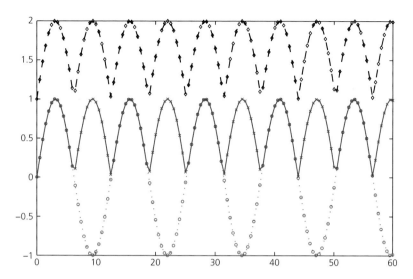

In the previous figure, we shifted one of our variables upwards by adding a constant value of 1.

One important aspect of figures is to be descriptive, and for that we can add labels on the axes, titles and legends. We can do that with the following commands: 'xlabel', 'ylabel' and 'zlabel' to add labels to each of the coordinate axis. In this case we do not have a z-axis. Titles are inserted with the command 'title' and legends with 'legend' like this:

```
xlabel('Time of the experiment [minutes]','fontsize',14);
ylabel('Values  [arbitrary units]','fontsize',16);
title('Figure one','fontsize',18);
legend('Original value', 'Absolute values',...
       'Absolute plus a constant');
```

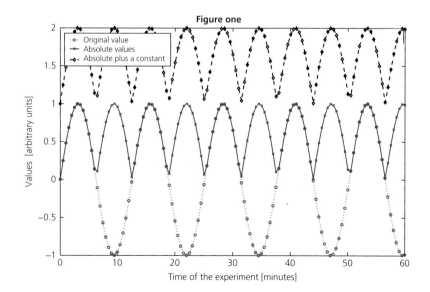

Notice that it is possible to have different font sizes for each of the labels that we add to the image. It should be clear that the commands in Matlab have names that should be fairly intuitive to use. If we want to add a grid we use the command 'grid' and turn it 'on' like this:

```
grid on;
```

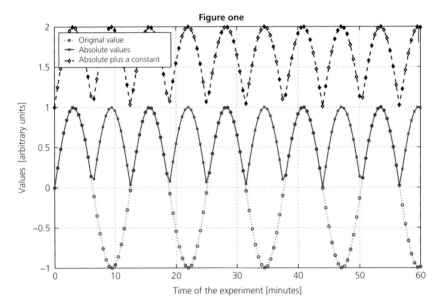

To remove the grid, the command has to be followed by 'off' like this:

```
grid off
```

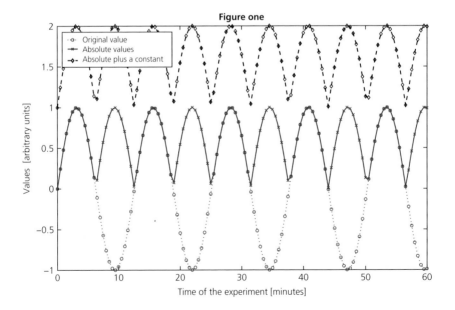

1.7   Random Numbers

Matlab has many functions that make it is easy to generate and manipulate data sets with a few simple commands. Some of the most useful functions to generate data are those that create matrices with zeros, ones or random values:

```
matrix_7          = zeros (1,2);
matrix_8          = ones  (3,4);
matrix_9          = rand  (15,16,25);
matrix_10         = randn (7,8,9,10);
```

   Notice how it is possible to create matrices of any number of dimensions just by addressing the dimension itself within the corresponding size. The previous lines have created: (a) a matrix with two zeros in a single row, (b) a matrix with ones in three rows and four columns, (c) a 3D matrix with random numbers distributed uniformly between 0 and 1 with 15 rows, 16 columns and 25 levels and (d) a 4D matrix with Normally distributed random numbers.

   It is important to highlight that random numbers come in many different forms, or more technically speaking, follow different probability distributions. The Normal or Gaussian distribution is very common in statistics, and is specified in Matlab with the command 'randn'. To obtain random numbers that follow a uniform distribution we use the command 'rand' to generate a series of numbers that have equal probability of occurrence between 0 and 1. To visualise the difference between these two random number distributions, we can use the command 'hist' to display the histogram of each matrix. The histogram is a well-known statistical graphical representation of the distribution of data, which divides the range of a number of values into a series of "bins". The values are then assigned to a specific bin and the number of values per bin is counted. Each bin is then displayed with a height proportional to the number of elements it contains. To display the histogram of the elements of a matrix, we can type:

```
hist(matrix_9(:),50)
```

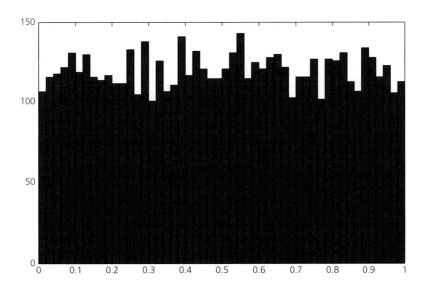

and this generates a histogram with 50 equally spaced bins. Notice that the colon operator has been used once more. In this case it is necessary as the matrix is three dimensional, and we are only interested in the histogram of all values of the matrix regardless of their position within the matrix. Similarly we can visualise the histogram with the Normal distribution:

```
hist(matrix_10(:),50)
```

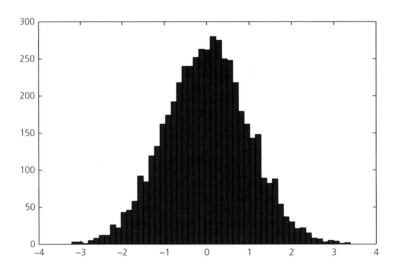

For completeness, we will present a third distribution of random numbers, the Poisson distribution. Applications of this distribution can be found in areas as diverse as telecommunications, astronomy, radioactivity and biology. The command to generate random numbers with a Poisson distribution is 'poissrnd'. Its first argument is a parameter called "lambda" related to the shape of the distribution and then the size of the distribution. For a distribution with lambda=3 and a matrix size 200×300 we type:

```
matrix_11 = poissrnd(3,2000,300);
```

Before we display the distribution we will save the values and positions of the bins into new variables for two reasons. First, it may be interesting to use these values later on, and second, we can plot the values in different ways. For the time being, we will plot the distribution with bars in the same way as before with the command 'bar', which works similarly to 'plot'. The first argument corresponds to the horizontal axis, the second to the vertical axis and a third can define the plot colours:

```
[values_poisson,bins_poisson]=hist(matrix_11(:),15);
bar(bins_poisson,values_poisson,'r')
```

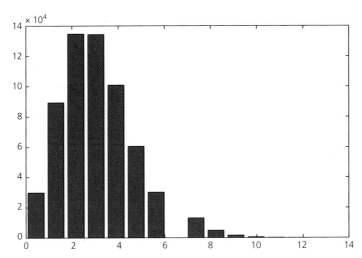

We can also display each of the values as a single line that is terminated with a circle marker with a command called 'stem'. Every time that we use a graphical command like 'plot' or 'hist', whatever is displayed on the current figure is erased to have a clean figure. We can indicate to Matlab that it should not erase the information previously displayed by using the command 'hold' like this:

```
stem(bins_poisson,values_poisson,'markersize',10,...
    'linewidth',3)
hold
plot(bins_poisson,values_poisson,'r:','linewidth',2)
```

Current plot held

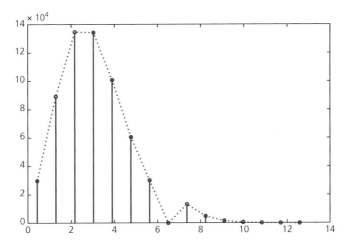

Notice how the distribution is displayed in two ways, with the dotted line between the points and with the "stems" with a solid line. Notice also how we can manipulate the parameters that define the lines; we can change the width of the line by specifying the parameter 'linewidth', which has a default value of 0.5, and the size of the marker, which has a default value of 5.

1.8   Statistics in Matlab

Some of the most useful functions of Matlab are those related with the calculation of statistical measurements like the mean or standard deviation. As in the previous cases, it is important to distinguish if you want to obtain a single measurement, the mean for instance, for all the values of a matrix, or alternatively, you are interested in obtaining mean values per column, per row, per level, etc. If we are interested in the arithmetic mean of all the values of a matrix we can use the colon operator to rearrange the matrix into a single column and then calculate the mean value in the following way:

```
mean_matrix_9              = mean(matrix_9(:))
```

```
       mean_matrix_9 =

          0.5006
```

The previous command calculates the average value of the elements of the 'matrix_9' and stores them in a new variable called 'mean_matrix_9'. If we use the function 'mean', without the colon operator, Matlab will calculate the mean for each individual column of the matrix, and thus the result would be a matrix of 1 row, 16 columns and 25 levels. We should recall that the original matrix had dimensions (15,16,25). We can then calculate the mean value of each dimension of a matrix. To do this we need to indicate the dimension of interest like this:

```
mean_matrix_column      = mean(matrix_9,1);
mean_matrix_row         = mean(matrix_9,2);
mean_matrix_level       = mean(matrix_9,3);
```

It is possible to pass the output of a function as an input argument of another function and thus concatenate the effect of the functions. For example, we can concatenate two commands 'mean' in a single instruction like this: 'mean(mean(matrix_9))'.

The innermost 'mean' function calculates the average of the columns. The second function calculates the average per row, as the columns will only have one element and you cannot calculate a average of a single element. To calculate the average values over the third dimension, the levels, we can use a third 'mean' function like this:

```
mean(mean(mean(matrix_9)))
```

```
       ans =

          0.5006
```

As 'matrix_9' has three dimensions, this last instruction has calculated the average value of all the elements and it is equivalent to 'mean(matrix_9(:))'.

In the same way the average value is calculated, other statistical measurements are easily computed. In some cases, it may be useful to store these values in a single variable, which can be saved and later compared against the other measurements obtained from a different process. We can easily combine different values by placing them in between square brackets to form a new matrix like this:

```
stats_observation_1 =
    [ mean(observation_1) min(observation_1) ...
      max(observation_1); median(observation_1)...
      sum(observation_1) std(observation_1)]
```

```
    stats_observation_1 =

       0.0237    -1.0000     0.9989
       0.0243     2.8713     0.7113
```

Notice that again we wrote the instruction over several lines by using three consecutive dots ("...").

So far, we have calculated the mean, median, minimum and maximum values, the sum of all elements and the standard deviation and stored them in a single variable called 'stats_observation_1'. There are some other statistical calculations whose result is not a single value, but rather a series of values. Two of the most useful ones are the commands 'sort' to arrange the elements of a vector in ascending or descending order, and 'cumsum', which performs the cumulative sum of the elements in a matrix. We can visualise these commands with the previously created variables of observation_1. Again we will sort the elements in the same instruction as the plot:

```
figure
plot(time_range,observation_1,'b-o',time_range,...
     sort(observation_1),'r--x');
```

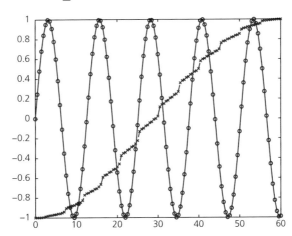

The sorted values of the sine function span the range from −1 to +1 as expected. What is probably more interesting to observe is that the increments between the values are not perfectly uniform, but rather have "jumps" that separate some regions where the increments are smaller.

The cumulative sum is useful in different contexts. As a first example, consider that the values of the matrix 'observation_2' correspond to an experiment and we want to observe, not only the actual value, but the behaviour with time of those values, that is, the values may look random, but could have a trend to increase or decrease with time. If we calculate the cumulative sum, and the sample is sufficiently large, the final values of the cumulative sum should tend to 0. If the values on the other hand are not completely random, there will be a trend, positive or negative of the values. It may also be the case that over a certain period the values may tend to be more positive than negative and the cumulative plot will show this with a line above 0. In these cases, it is useful to place a grid on the figure to denote the values of both axes.

```
plot(time_range,observation_2,'b-o',time_range,...
    cumsum(observation_2),'r--x');
grid on
```

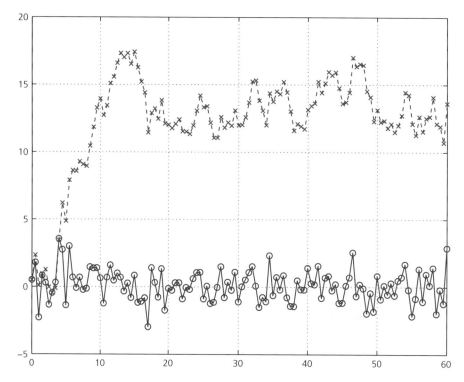

A second application of the cumulative sum is to transform a probability distribution like the Gaussian or Poisson, to a cumulative distribution. To illustrate this, we can display the bar chart that we previously showed for the Poisson distribution, but this time we will combine it with its cumulative distribution. In addition, we will normalise the values by the sum of all the values. In this way the distribution will add to 1 and the last value of the cumulative distribution will be again 1.

```
figure
hold on
bar(bins_poisson,values_poisson/sum(values_poisson),'r')
```

```
plot(bins_poisson,cumsum(values_poisson)/sum(values_poisson),...
    'b-o')
grid on
```

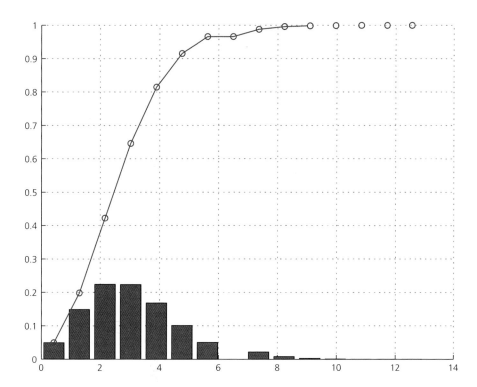

Now we can see how the cumulative distribution behaves: the first five bins of the distribution contain more than 80% of its values, and the first seven reach above 95%. From that point forward, the distribution reaches its maximum value with very small increments.

Notice that we created a new figure to display the two graphs together. As was mentioned previously, the default condition of a Matlab figure is that every time that a graph is generated, it erases whatever it has previously displayed and shows the next graph. In this case, we indicate to Matlab that the figure should 'hold on' to the graphs until indicated otherwise, which is done with the command 'hold off'.

1.9   Displaying Two-Dimensional Matrices

The data that we have displayed so far is one dimensional, as each of the matrices has a series of values in a single dimension and these values are placed in a two-dimensional plane in which one of the axis (the vertical) corresponds to the values of one matrix, and the other axis (the horizontal) corresponds to the values of the other matrix. If we only passed one matrix as an argument to the display command the horizontal axis corresponds to the order at which the values appear in the matrix. We have shown only 'plot', 'bar' and 'stem', but there are many others like 'rose', 'feather', 'compass' or 'stairs'. We recommend that you read the 'help' pages of these commands to familiarise yourself with these different options of displaying data.

Two-dimensional matrices will be displayed in a 3D plane, in which the values of the elements correspond to the position of one axis (the vertical) and the position of each element in the matrix will correspond to two axes in a horizontal plane. For instance to display the values of 'matrix_4' so that the values of the matrix correspond to the elevation of points of a grid, we use the command 'mesh' and pass as a single argument the matrix itself like this:

```
figure
mesh(matrix_4)
```

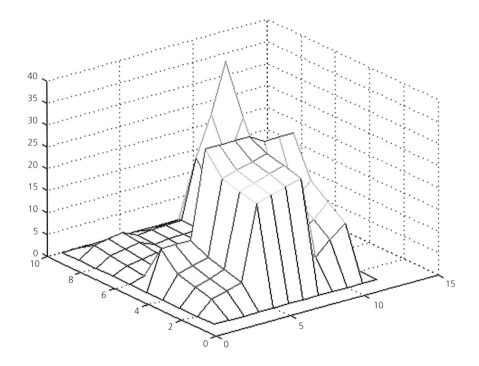

The grid describes the values of 'matrix_4' as an altitude in the vertical axis and adds lines that connect each point. The colours of the lines are related to the magnitude of the value (altitude) itself. It is easy to display this as a colour bar on the same figure with the command 'colorbar':

```
colorbar
```

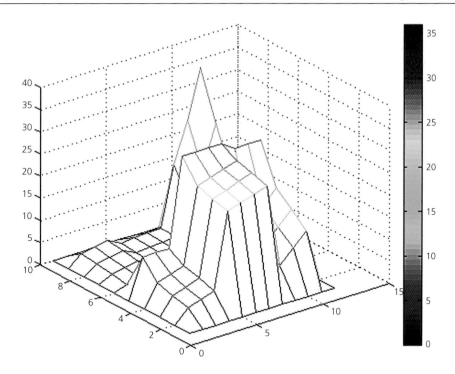

The colorbar can be removed by passing the argument 'off' like this:

```
colorbar('off')
```

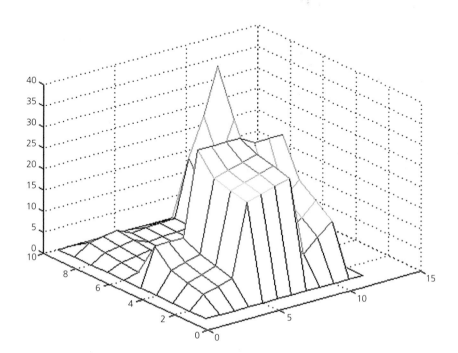

We will explore now another useful 3D display in which the regions between the points are coloured with surfaces; the command to generate this kind of plot is 'surf', short for surface:

```
surf(matrix_4)
axis tight
```

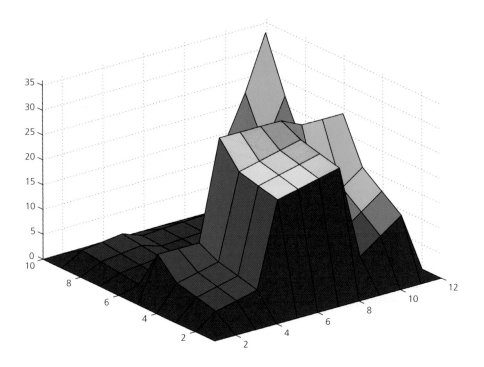

Notice that after displaying the surface we used the command 'axis', which is used to modify the conditions of the axes of the figure. In this case we indicated that we wanted to fit the surface "tightly" into the axes by typing 'tight' immediately after 'axis'.

The default display of the command 'surf' is to draw black lines between the values of the matrix and surfaces with different colours; the colours are relative to the altitudes of the surfaces. It is possible to modify the display of the surfaces very easily by changing the "shading". For example, this command removes the lines:

```
shading flat
```

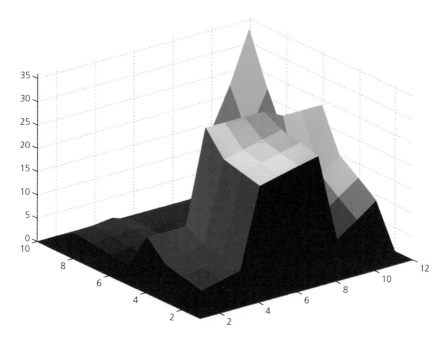

and this command will average the colours (the technical term is "interpolation" of values) within each surface so that the colours do not "jump" between values:

```
shading interp
```

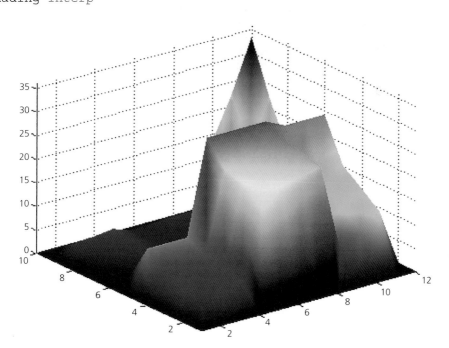

At this moment, you may be interested in visualising the surface from a different point of view, as there may be important information that is not visible from the current position. To change the view angle you can use the command 'view', which will change the view defined by an "angle" of rotation around the surface and the "azimuth" (moving to a higher or lower position). We indicate these values between brackets after the command like this:

```
view(30,20);   %  %Change angle of the surface
```

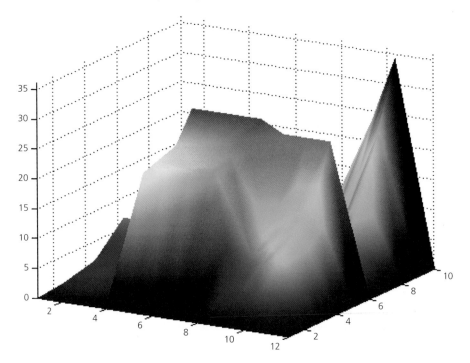

The angle varies between 0 and 360; in this case we selected 30. The azimuth varies between −90 and 90, with 20 in the present example.

The previous instruction changed the angle of view of the surface and also introduced some comments. In Matlab, anything that is written after the per cent symbol (%) is considered to be a comment. It is ignored by Matlab and does not create any output or function. The comments are very useful when writing larger functions or code, as it is good programming practice to document the codes with sensible comments and explanations that may be useful to other people reading the code. Another important reason to include comments and use sensible names for the variables, is that after some time of not reading a certain piece of code, it is easy to forget what we were trying to do with a certain line or what the variable 'q' was used for.

To visualise a surface from many different angles it could be possible to type the command 'view' many times to change the angle, but this would be slow and tiring. Matlab allows you to use the mouse to rotate the figure directly. You need to enable the rotation of the figure, either by clicking on the rotation angle of the figure or by typing:

```
rotate3d on
```

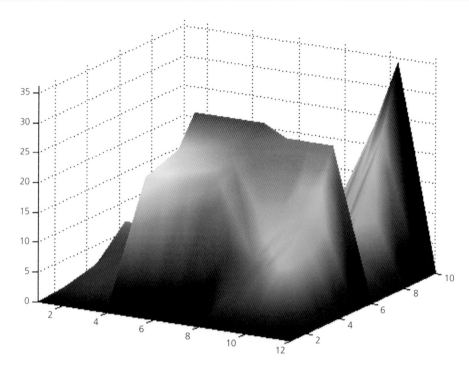

and then you can rotate your figure with the mouse. Place the mouse on top of the surface, press the left button and rotate the surface at will. When you release the button the surface will stop rotating.

Two important concepts in the generation of figures follow: (1) you can open many figures and (2) you can place several plots in a single figure. This latter option will save you having to export your figures separately and then merging and annotating them with a separate software. Matlab allows you to do everything in one go!

To open a new figure you simply type the command 'figure'. Each time you type 'figure', Matlab will open a new window to display the figure. When you have several figures open, it is useful to address one specific figure and that is done by passing the figure number as an argument to the command 'figure'. If the figure with that number does not exist yet, Matlab will create a new figure. If the figure does exist, it will be brought to the foreground and all the instructions that follow will be placed on that figure.

In Matlab it is very easy to partition a figure so that you can display several graphs in a single figure. This is useful when we want to create figures to be submitted to journals. Chapter 8 has an entire section on creation of high quality figures, but we will introduce some of the basics here.

The command 'subplot' will partition a figure into several sections. The command 'subplot' can be used in different ways. The most common is to specify a number of partitions, rows and columns, and address each of them to place a graph in each. This is done through the input arguments (row, column, position). For example, to create one figure with six graphs, in two rows and three columns, you will create each one of them like this:

```
subplot(2,3,1); mesh(matrix_4);   axis tight;
```

```
subplot(2,3,2); mesh(matrix_4);   axis tight; view(20,30)
subplot(2,3,3); mesh(matrix_4);   axis tight; view(30,60)
subplot(2,3,4); surf(-matrix_4); axis tight;
subplot(2,3,5); surf(-matrix_4); axis tight; view(20,30)
subplot(2,3,6); surf(-matrix_4); axis tight; view(30,60)
```

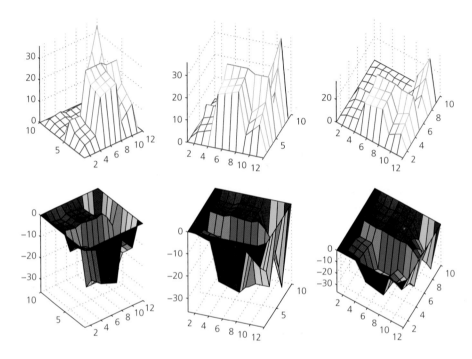

Notice how several Matlab commands were written in a single line. This helps when you want to repeat similar commands from the Command Window as Matlab provides a very useful tool for this. Once you have typed something into the Command Window, it is stored in the 'command history', which is a list of the most recent commands you have typed. The size of the list exceeds several hundred commands, so you can go back many days! To retrieve a command you only need to place the cursor on the Command Window (place the mouse on top of it and click once) and move the "up" arrow to go back in time and "down" to go forward. If the command was typed relatively recently it is easy to click the arrow key several times to find the command. A shortcut for this is to type the first letter of the command, say "v", and then when you press the arrow up, it will retrieve only those commands that start with "v", like 'view'. Once you have found the command you want, you can modify some parameters, such as the view angle for our example, and press enter to execute the instructions again.

There is another way to use 'subplot' by specifying a position, in coordinates, of the graph within the figure. This will not be covered here but in Chapter 8.

To erase all the content of a figure, we use the command 'clf', which stands for "clear figure". If we only want to erase the content of one subplot we use the command 'cla', which stands for "clear axis".

1.10 Scripts, Functions and Shortcuts

Among other useful things of Matlab (such as powerful graphics, numerous pre-defined functions, etc.) are the options to "make things easier". More specifically, to make repetitive actions easier.

Once the analysis of a given problem begins to become complicated, it will be necessary to execute several commands in sequence: `'read data, pre-process, measure, store'`. It would be cumbersome and error-prone to be typing them again and again, even with the use of the up/down arrows to retrieve commands. For this kind of task, which requires a series of commands, few or many, it is better to save them either as a separate file (called m-files as the extension is .m) or as a `'shortcut'`. We begin with `'shortcuts'`, as these are absolute time-savers. You can define as many shortcuts as you want and then use them with a simple click. In this section, three useful examples are provided, but the important part is to learn how to create them and then you can add as many as you want.

To add a shortcut, you need first to have the code for a repetitive task, for example, to export a JPG or TIFF image from the graphics in a figure. Second, you need to know where shortcuts are placed and used. For older version of Matlab (like R2010), at the top of the Command Window there is a line with "MATLAB" and the version you are running, then another line with "File", "Edit", etc., then another line with the icons for "New File", "Open", "Copy and Paste", etc., and below that there are the lines for shortcuts. For more recent versions of Matlab (like R2013), at the top of the window there are several "tabs": "Home", "Plots", "Apps" and "Shortcuts".

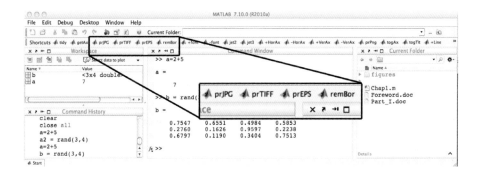

For version R2010, if you have not added any shortcuts, then the line will only have "How to Add" and "What's New". For version R2013, there will be "New Shortcut", "Organise Shortcuts" and "Quick Access".

We will now introduce three useful shortcuts, which are very handy when printing your figures, either for publication (high quality TIFFs) or web-based dissemination (medium quality JPEGs). The first two shortcuts will export an image in TIFF/JPG format when you click on them. The third one toggles the axis and the borders of an image.

For the purpose of these examples, we will generate a simple plot in a figure similarly to the previous examples.

```
figure
plot(randn(1,100),'b-x')
```

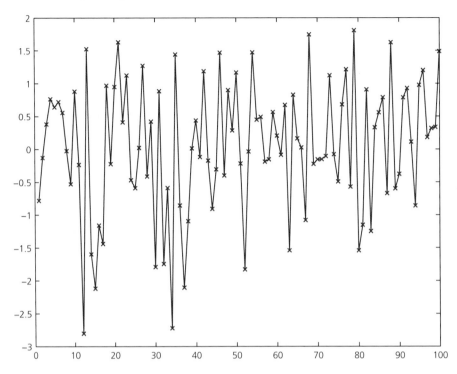

To export the figure to an image file, you can use the 'print' command, which will create the file according to the parameters you use to define it. For example, to define the resolution in dots-per-inch (dpi), say 200, you use the parameter '-r200', and to specify that you want a JPG image you use the parameter '-djpeg'. The final parameter that is required is the name of the file you want to create. We will consider that the name will be stored inside the variable 'filename' and that can be typed in the Command Window before you click on the shortcut. The shortcut will prompt for a name if the variable 'filename' does not exist. As mentioned previously, try to use names that are explicit and clear, using uppercase letters can help the names be more readable, for instance:

```
filename                = 'myFirstFigure.jpg';
```

The following code will be used for the shortcut:

```
gcf;
if ~exist('filename','var')
    filename=input('Please type the name of file');
end
print('-djpeg','-r200',filename)
```

The first line "grabs" the figure with the command 'gcf', which stands for "get current figure". The second line checks if a variable called 'filename' exists and if it does not, it prompts the user to type a name, which will be stored with the command 'input' in the variable 'filename'. Then the figure is exported with the command 'print' as a JPG figure with a resolution of 200 dpis.

For version R2013 you need to click on "New Shortcut", whilst for version R2010, you need to right-click with the mouse on the line of shortcuts. This will open a window, then select "New Shortcut". Select a name for your shortcut and type it in the "Label" box. In the "Callback" box, type (or copy and paste) the lines of code that define the shortcut and click "save". The figure below shows the window for the shortcut to remove borders described below.

The name of the shortcut should be easily identified with the function it performs. It is also a good idea to keep the name short, as you may add many shortcuts later and you do not want to use a lot of space with a long name. For example, 'PrJPG' could be used for printing JPG images.

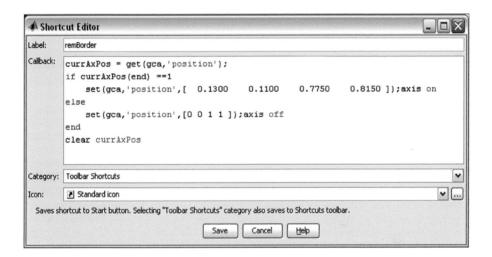

You can select the type of icon in that same window. Immediately after the window is closed, the new shortcut will appear in your Matlab Command Window. To use it, you will first generate a figure as described above, and then write down (at the Command Window) the name with which you want to save the figure and then click the shortcut. Immediately after you click the shortcut, a file with the name you specified will appear in the same folder as you are working, e.g. something like "C:\Documents and Settings". If you select a very high resolution image, it may take a few seconds more to generate the image. You can now view your image with your favourite graphics software by double-clicking on it. Once you have the shortcut enabled, you can export a new figure by changing 'filename' and clicking the shortcut button; it could not be simpler! Remember to change the name of the file; otherwise you will be overwriting your previous images.

If you want to save the images in a different folder, you can change folders by clicking the icon at the very right of the line with the current folder (R2010 three dots, R2013 an arrow pointing down) as shown in the image below.

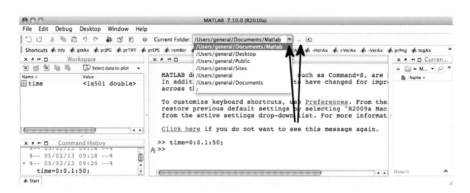

You can also change the folders from the Command Window with the command 'cd', which stands for "change directory", as it is done in Unix and Linux operating systems. If you want to create a new folder (or directory), you can do with the command 'mkdir', to "make a directory" followed by the name of the new directory, like this:

```
mkdir imageResults
```

This will create a new folder called 'imageResults'. Once you have the new folder, you can save your results into that folder by modifying 'filename' so that the folder and the name of the file are both specified like this:

```
filename              = 'imageResults/myFirstFigure.jpg';
```

Then, when you click the shortcut, the image will be saved, not in the current folder but in the folder 'imageResults'. As a general rule, it is better NOT to use spaces or special characters (!, @, $, %, ^, *) as part of the names of files or folders as not all computer systems interpret them the same way. Keeping names simple and descriptive with a-z, A-Z and numbers only is a good practice. If necessary you can use underscore "_".

When you want to create high-quality images for submission into journals you will require a format different than JPG. You can easily create TIFF, EPS or other image formats in Matlab by modifying the parameters you use with the command 'print'. The name that you select should also finish with 'tiff' or 'tif', like this:

```
filename              = 'imageResults/myFirstFigure.tif';
```

The new shorcut (for instance 'PrTIFF'), is going to be very similar to the previous shortcut with the exception of the print command, in this case to save a high quality 'tiff' we will use the command with the following parameters:

```
print('-dtiff','-r400',filename)
```

Notice that in this case the resolution will be 400 dots per inch.

When you plot something in Matlab, the figure will contain by default the axis that shows the dimensions of the plot. This may be useful in some cases as seen in the previous examples, but in other cases, like images, you may want the image to fill the figure without any added axis. The following code can be used to create a shortcut that will remove the border that surrounds your plot, or add it back if it had been removed. A single click on the shortcut can tailor your image before printing it. Look at the difference between a figure with and without axes. First we can create the image like this:

```
figure
surf(-matrix_4); axis tight; view(0,90)
```

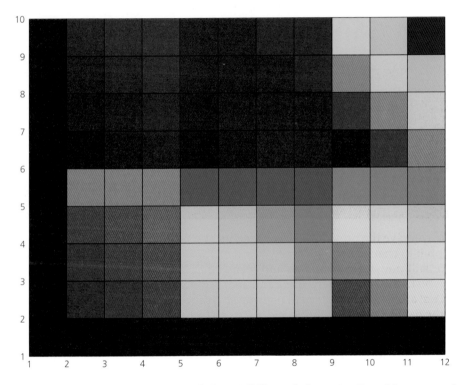

Next we toggle the position and the visibility of the axis. For this we need to determine the position. We can do this by saving the position of the current axis (with the command 'gca') into a variable ('currAxPos'). If the last parameter of the position, the height of the axis is equal to 1, the axis has been removed and should be included, otherwise, it should be removed and the figure expanded.

```
currAxPos = get(gca,'position');
if currAxPos(end) ==1
    set(gca,'position',[ 0.1300    0.1100    0.7750    0.8150 ]);
    axis on
else
```

```
    set(gca,'position',[0 0 1 1 ]);
    axis off
end
clear currAxPos
```

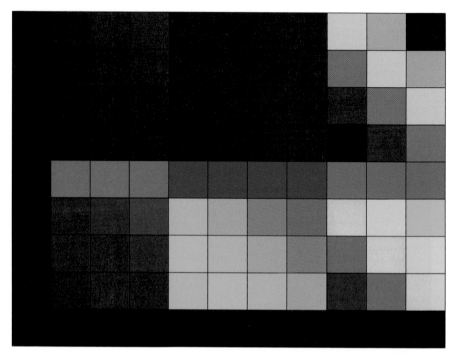

We can save the previous lines of code for the shortcut "ToggleAxis". We can then click on it to change the figure.

We will now leave shortcuts and focus on 'm-files'. Scripts and functions, also called 'm-files', are widely used and make life easier while using Matlab. These 'm-files' are normal text files that group several commands with the extension '.m', thus the name. Once these are saved, they can be used as any normal Matlab function or to execute all the commands of the file in sequence.

There are several ways to create a new script. One simple way of opening a new script on the editor is to press control-N (or command-N in Macs), which will open a new untitled file in the Editor. The Editor is similar to other text-editing software, but it contains numerous capabilities that are specific to Matlab.

Scripts are files that contain any number of lines with Matlab commands. You can write as many as necessary, for example, all the commands to generate a time variable, an observation, modifications to that observation with other Matlab functions and a plot of the results as described earlier. Once all the commands have been written down, you only need to save the 'm-file' with a descriptive name in one folder of your choice. Afterwards, whenever you want to repeat the commands, you can open the m-file again and run your commands, with the same or different parameters.

To run the commands in the m-file script, you can simply type the name of the m-file on the Command Window and press enter, or you can 'run' the script from the

Editor by clicking the "run icon", a green triangle over a white square or by pressing Alt-Control-R (Alt-Command-R in Macs). This will execute all the functions on the script. Note that you need to be working on the same folder ("Current Folder" on the top of the Command Window) as the folder where you saved your script.

Functions are similar to scripts in the sense that they are 'm-files' that have any number of Matlab commands, however they differ in that they may have input and output arguments that may be used as parameters of the function. The first line of a code of a function is what makes it different from a script. Whilst a script can begin with any valid Matlab instruction, a function must begin with the keyword 'function', followed by the output arguments (if any), the name of the function, which should be identical to the name of the 'm-file', and the input arguments (if any). If we think of a certain test function that receives three input arguments and then calculates two output arguments, the 'm-file' 'testFunction.m' would begin with a line like this: 'function [output1,output2] = testFunction (input1, input2, input3)'.

Unlike a normal Matlab instruction line that needs to be finished with a semicolon (;) to prevent the echo to the screen, the function line does not require it to be terminated with the semicolon. After this first line, we can write as many instructions as necessary to calculate the output arguments from the input arguments. As it was mentioned, neither input nor output are strictly necessary. For example, a function may receive an input and display it, without any output, or another function can read pre-defined data, like the values of the stock exchange, and calculate something.

1.11 Using Help

Matlab has extensive help pages. To get help over any command you can type help and the command, for example:

```
help sum
```

```
SUM Sum of elements.
   S = SUM(X) is the sum of the elements of the vector X. If
   X is a matrix, S is a row vector with the sum over each
   column. For N-D arrays, SUM(X) operates along the first
   non-singleton dimension.
   If X is floating point, that is double or single, S is
   accumulated natively, that is in the same class as X,
   and S has the same class as X. If X is not floating point,
   S is accumulated in double and S has class double.

   S = SUM(X,DIM) sums along the dimension DIM.

   S = SUM(X,'double') and S = SUM(X,DIM,'double') accumulate
   S in double and S has class double, even if X is single.

   S = SUM(X,'native') and S = SUM(X,DIM,'native') accumulate
   S natively and S has the same class as X.
```

```
Examples:
If   X = [0 1 2
          3 4 5]

then sum(X,1) is [3 5 7] and sum(X,2) is [ 3
                                          12];

If X = int8(1:20) then sum(X) accumulates in double and
the result is double(210) while sum(X,'native')
accumulates in int8, but overflows and saturates
to int8(127).

See also PROD, CUMSUM, DIFF, ACCUMARRAY, ISFLOAT.

Overloaded methods:
    codistributed/sum
    gpuArray/sum
    sym/sum
    timeseries/sum

Reference page in Help browser
    doc sum
```

or you can open the help window by clicking the blue question mark icon in the Command Window and then searching for your command of interest. The web site of Mathworks also contains on-line documentation for all Matlab functions. Moreover, there are discussion forums, blogs, videos, file exchanges with excellent resources. Examples and answers to users' questions are posted. The web site is http://www.mathworks.co.uk/matlabcentral/

It is strongly recommended that you spend some time exploring the help pages of Matlab.

CHAPTER 2

Introduction to Images

2.1   An Image as a Matrix

The simplest way of thinking of an image is that it is a grid of points or elements and each of them will contain a certain value. The value of the element is related to the colour or intensity of the image itself. In the technical jargon, the grid is called a "matrix", and each point of the grid is called a "pixel", which is the combination of the words "picture element". The images we will refer to along this text will all be digital images, and the size of the grid will define the resolution of the image. When data sets contain volumetric information, like those acquired from Magnetic Resonance Images or multiphoton microscopes, the elements are related to a volume and thus the "volume elements" are called "voxels". There is no specific term to describe the elements of data sets with higher dimensions.

The size of an image varies widely, for example, it may be a square of 128×128, 256×256 or 512×512 pixels or it may depend on the resolution of the cameras used to capture the image. In the general case an image can be of size m-by-n, 'm' pixels in the rows or vertical direction and 'n' pixels in the columns or horizontal direction. The value or intensity of the pixels can be related to light intensity, attenuation, depth information, concentration of a fluorophore, or intensity of radio wave depending on the type of acquisition process through which the image was captured.

This intensity level of each pixel is the essential information that forms the image. The intensity values of the pixels are also used to perform operations like segmentation or filtering. In addition, we can use the intensity to extract useful information, like number of cells in an image. In other cases, the relationship between the intensities of neighbouring pixels and their relative variation may be of interest, as it is in texture analysis.

The intensity level of a grey scale image describes how bright or dark the pixel at the corresponding position should be displayed. There are several ways in which the intensity of each pixel is recorded. For example, a pixel may have one of two options, 0 or 1, where 0 will be displayed as low intensity (or black in some cases) and 1 will be displayed as high intensity (or white). These two options correspond to what is called in mathematical terms as 'logical' or 'Boolean' data types.

Another very common data type is called 'uint8' (un-signed integer of 8 bits), which assigns an integer value (no decimals) between 0 and 255 to represent the

Biomedical Image Analysis Recipes in MATLAB®: For Life Scientists and Engineers, First Edition.
Constantino Carlos Reyes-Aldasoro.
© 2015 John Wiley & Sons, Ltd. Published 2015 by John Wiley & Sons, Ltd.

brightness of a pixel. The value 0 corresponds to black and 255 to white. Most images imported to Matlab from digital cameras are stored with this data type. Some images, especially those that have been acquired with high quality imaging equipment may use a data type called 'uint16' (un-signed integer of 16 bits). Like uint8, these assign integer values to represent the intensity of the pixels, however they use 16 bits for the resolution, thus the range is far larger with options between 0 and 65,535. The maximum possible value is determined by the number of bits, i.e. $2^{16} = 65,536$, likewise $2^8 = 256$. (Notice that the options begin at 0, thus the maximum is equal to $2^n - 1$ where n is the number of bits.)

Another important data type, which is the one that Matlab uses by default when a new variable is created, is 'double'. The 'double' data type is a double precision (there exists also a single precision less frequently used), which allows decimal and negative values, and for very small and very large values, it uses scientific or exponential notation. For example, 1,200 can also be written as "$1.2e + 03$" and 0.0001 is equivalent to "$1.0e - 04$". The importance of a 'double' is that integer values are limited in scope and problematic when used in mathematical operations; therefore it is recommended to use variable of data type 'double'.

Colour images contain important information about the perceptual phenomenon of colour related to the different wavelengths of visible electromagnetic spectrum. In many cases, the information is divided into three primary components, Red, Green and Blue (RGB), or psychological qualities such as hue, saturation and intensity. Prior knowledge of the object's colour can lead to classification of pixels, but this is not always known. In Matlab, a 2D matrix is assumed to be a grey-scale image, and a 3D matrix with any number of rows and columns and three levels is assumed to be an RGB colour image. We will soon learn that two-dimensional images can be displayed with different "colour maps", which are not related to the RGB components of a colour image.

2.2 Reading Images

To begin our introduction to images, we need an image. Matlab allows many ways of importing images into Matlab. The simplest one is to use the mouse or pointing device to "drag-and-drop" an image file, either from the desktop or a specific folder in your computer into the Matlab Command Window. In this way, you can navigate to find your image and drop it into the Command Window.

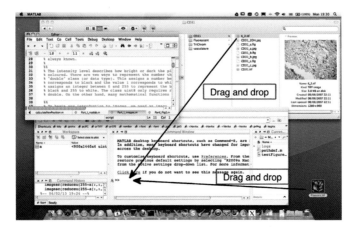

When you "drop" the image in the Command Window, a dialogue window called "Import Wizard" will appear asking for a confirmation of the input into Matlab. You only need to click "Finish" or press Enter and the image will have been imported into Matlab. You will see that the Command Window will show you the actual command that was prompted by the drag and drop action. It will be something like this: `'uiopen('/Users/general/Desktop/Fluores.tif',1)'`.

The process of drag-and-drop is using the command `'uiopen'` (user-interface open), which is one of many commands that employ "user interfaces" or windows that allow the users to interact with Matlab. The command then takes as an input parameter the location of the image. Notice that the command has written the `'path'`, that is, all the folders and subfolders where the image is stored, in this case `'/Users/general/Desktop/'`. In Windows it would be something like `'C:\Documents and Settings\...'`, besides the name of the image file: `'Fluores.tif'`. When you have many images to process, the path may be used as part of a "loop" to read the images. Alternatively, you may want to read several images that are located in different paths. In those cases, you need to know the location of the image and specify that path in the command to read the image.

To read an image without the use of the mouse, you can use the command `'imread'`. This command will have as input parameters the path to the image itself, which can be the same as the one to the image we previously imported through the drag-and-drop process. In addition, `'imread'` has an output parameter, that is, it can assign the outcome of its process to any variable that we want. For example, we may want to assign a particular name to our image like this: `'fluorescent_cells = imread('/Users/general/Documents/Matlab/Fluores.tif');`.

In this case, we have provided the name and path of the image we want to read to the `'imread'` command and instructed Matlab to store that image in a variable called `'fluorescent_cells'`.

Typing long paths can be cumbersome and prone to typing errors, and there are two ways in which you can simplify your life in Matlab. The first is to change folders or directories in Matlab. In this way, you can store all the images of one experiment in a folder and then do all the processing in that folder. You can change the current folder that is specified in the top of the Matlab Desktop window by typing the path of the desired folder on that line. If you click on the arrow at the end of that line, Matlab will show you all the folders you have been working on in the past, and if you click on the icon with the three dots a window will pop up. Through this window you can select the folder where you have your data stored.

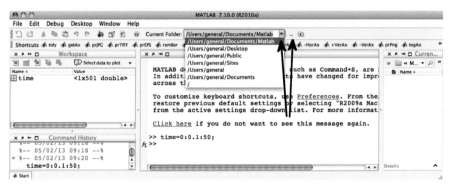

Once you are in the selected folder, you can simply read an image by typing its name: `'fluorescent_cells = imread('Fluores.tif');`. Notice that we have written the name of the image between singe quotes ('). This has indicated to Matlab that we are specifying a series of characters (a string). Without the quotes, Matlab would assume that whatever we write inside the brackets would be a variable and would return an error, unless of course, the variable would contain a string with the path and name to the image to be read.

When you have a large number of folders with data, or once you start saving your own functions, changing folders every time may be time consuming. There is another way to avoid typing the whole path each time we read an image and that is by adding a folder to the general Matlab set of "paths". Matlab has many files stored in different folders, so it stores the access to the files in a set of paths. You can store the path to the folders with your data by selecting from the main menu on the top left "FILE > SET PATH" in version R2010. For version R2013 you can find this on the "Home" tab, towards the right under "Environment". Setting a new path will open a window that will allow you to select the folder with your data or m-files and store it with other Matlab paths. You can select to add one folder, or, if you have nested folders with data or functions, add a folder and all the subfolders inside it. Once you have finished adding folders you should click "SAVE" and "CLOSE".

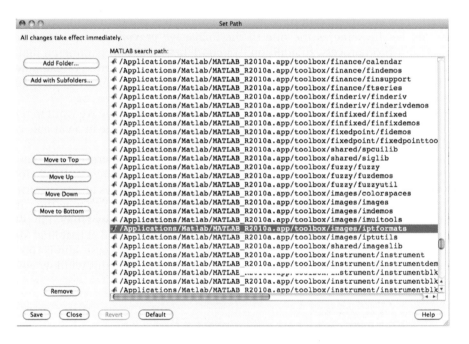

Once you have added a folder to the path, it will be stored for future sessions of Matlab. If you need to delete a folder, follow the same procedure, "FILE > SET PATH", and select the folder(s) to remove and click "SAVE".

The command 'imread' is quite powerful as it automatically checks the format in which the image was saved (jpeg, tiff, bitmap, etc.), which may or may not be compressed, and then assigns the values of the pixels into a two-dimensional or three-dimensional Matlab matrix. A very useful quality of 'imread' is that it can also read images that are located anywhere on the Internet (as long as they are not behind firewalls or password-protected). Instead of typing the path to the folder in a local computer, we can use the location of the image on a web site like this:

```
fluorescent_cells = imread(...
'http://caiman.group.shef.ac.uk/caiman/images/Fluores.png');
```

2.3   Displaying Images

With the previous line we have read an image and stored it in the variable called
'fluorescent_cells', which we can display with the command 'imshow' like this:

```
imshow(fluorescent_cells)
```

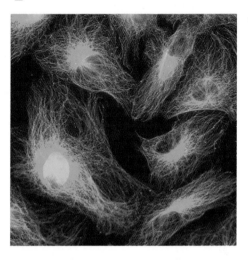

Image courtesy of Dr. Chryso Kanthou, The University of Sheffield, UK.

We can see that the image corresponds to a series of cells that have been fluorescently
labelled with the colours red, green and blue.

A useful source of images, which in many cases are released in the public
domain, freely licensed media file repository and "may be used, linked, or reproduced
without permission" is Wikipedia (http://www.wikipedia.org/) and Wikimedia
Commons (http://commons.wikimedia.org/wiki/Main_Page). For example
we can read an image of normal circulating human blood observed with a scanning
electron microscope (authors Wetzel and Schaefer, February 1982), like this:

```
blood_cells = imread(...
'http://upload.wikimedia.org/wikipedia/commons/8/82/SEM_blood_cells.jpg');
```

In the previous line, we instructed Matlab to read an image (imread) from
a web site (the Wikimedia Commons address) and store it in a local variable
(blood_cells). Before we display the image, we may be interested in knowing its
dimensions. We can find out using the command 'size', and we can store the results
in three variables, one for each dimension of interest:

```
[rows,columns,levels]              = size(blood_cells)

        rows =

    2239
```

```
columns =

    1800


levels =

    3
```

We can see that the image contains 2239 rows, 1800 columns and 3 levels. To display the image we use the command 'imshow'. In order to keep the previous image open, we will display the image in a new window or figure. We can do this by opening the figure before displaying it:

```
figure
imshow(blood_cells)
```

The image of the blood cells is a "grey-scale" image, as each pixel contains information related to a level of grey between 0 and 255. Besides grey-scale images, we may be interested in black and white images (purely black and white, without any intermediate greys), which are also called binary images, and colour images.

In some cases, it may be the case that the image is too big to fit in the screen and will be scaled down. This is one of the properties of the command 'imshow', which is special for images. We can also use the command 'imagesc' (image-scale), which provides more options for image analysis:

```
figure
imagesc(blood_cells)
```

There are some important differences between the commands 'imshow' and 'imagesc'. First, the axes along the rows and columns are normally shown with 'imagesc' whilst with 'imshow' they are hidden. Second, 'imshow' will scale automatically vertical and horizontal dimensions so that the pixels have the same dimensions in rows and columns. At the same time, it scales the size of the window proportionally to the image. On the other hand, 'imagesc' opens a standard size window and will adjust the size of the rows and columns to the size of the window, thus the pixels do not necessarily have the same dimensions in rows and columns. To further illustrate these differences, we can try the following command, which will plot only a quarter of the columns of the blood cells image:

```
figure
imagesc(blood_cells(:,1:end/4,:))
```

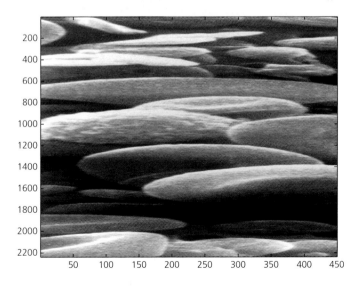

To keep the vertical and horizontal axis in the same proportion we can use the following command:

```
axis equal
```

And to remove the white space around the image we can use this command:

```
axis tight
```

If we want again to fill the image we simply change the axis with the following command:

```
axis normal
```

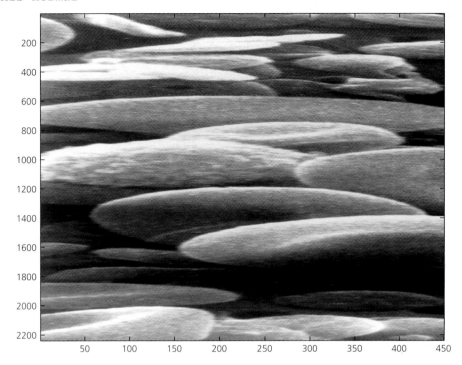

The windows in the screen can be modified in size by using the mouse and clicking on the edges or corners of the image and changing its dimensions (in some Macs you can only change by clicking on the bottom right corner).

There is another important use of the command 'axis'. In the figures that we have observed so far, the rows and columns begin in the top left corner, and grow downwards and to the right. In contrast, when we display signals, the coordinate system normally starts in the bottom left corner and grows upwards and to the right. In Matlab, the system where the origin (the place where the numbering starts) is at the bottom left and is considered the default Cartesian system (or Cartesian axes mode) and is obtained by writing the letters 'xy' after the command 'axis'. The alternative system with the origin at the top is obtained by writing the letters 'ij'; this system is called the "matrix axes mode". We can observe the two systems together in the following figure:

```
figure
subplot(121)
imagesc(blood_cells)
axis xy
subplot(122)
imagesc(blood_cells)
axis ij
```

2.4 Colormap

Some images only contain one colour channel, that is, their dimensions will be [rows × columns × 1]. In other applications, these images would be displayed as grey-scale images as there is no direct correspondence of a specific colour for each pixel. When these types of images are displayed in Matlab, there are several ways to display them. Some of these different options will artificially assign colours to the pixels according to some specific colouring rules called "maps" or in the context of Matlab 'colormaps'. In some cases, the maps can provide more contrast that the traditional grey scales. To illustrate these 'colormaps', we will first display only one channel of the fluorescent cells with grey scale:

```
figure
imagesc(fluorescent_cells(:,:,1))


colormap(gray)
colorbar
```

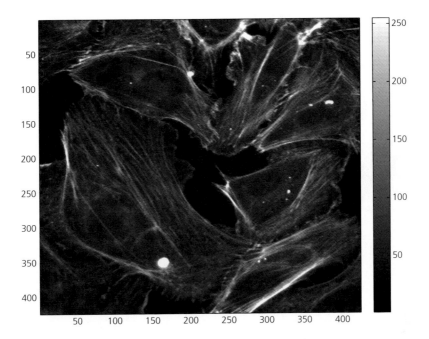

The command 'colormap' specifies the "map" we want to use for this image. In this case a grey-scale map is used, which is displayed in the colour bar on the right, which is added with the command 'colorbar'. To change the map, we simply type a different one, for example:

```
colormap(hot)
```

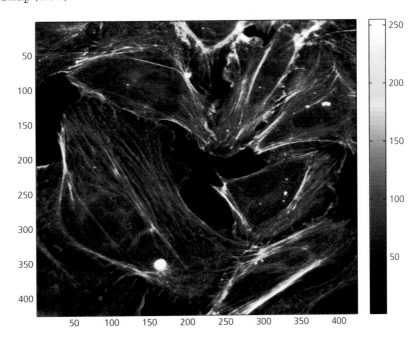

A very typical map, widely used due to its visual discrimination between low and high intensities, is the 'jet' map, which uses the colours of the rainbow.

```
colormap(jet)
```

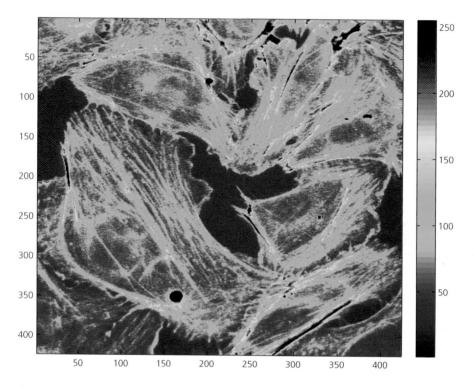

There are many different colormaps: 'hot, cool, bone, gray, spring, summer, autumn, winter, hsv, copper', etc. You might want to experiment with a few of these to see which is the best way to display your images. We can even manipulate the colormaps with mathematical functions. The colormaps are predefined matrices with numbers that assign weights for red, green and blue, for example.

```
jet
```

```
ans =

     0           0      0.5625
     0           0      0.6250
     0           0      0.6875
     0           0      0.7500
     0           0      0.8125
     0           0      0.8750
     0           0      0.9375
     0           0      1.0000
     0      0.0625      1.0000
```

0	0.1250	1.0000
0	0.1875	1.0000
0	0.2500	1.0000
0	0.3125	1.0000
0	0.3750	1.0000
0	0.4375	1.0000
0	0.5000	1.0000
0	0.5625	1.0000
0	0.6250	1.0000
0	0.6875	1.0000
0	0.7500	1.0000
0	0.8125	1.0000
0	0.8750	1.0000
0	0.9375	1.0000
0	1.0000	1.0000
0.0625	1.0000	0.9375
0.1250	1.0000	0.8750
0.1875	1.0000	0.8125
0.2500	1.0000	0.7500
0.3125	1.0000	0.6875
0.3750	1.0000	0.6250
0.4375	1.0000	0.5625
0.5000	1.0000	0.5000
0.5625	1.0000	0.4375
0.6250	1.0000	0.3750
0.6875	1.0000	0.3125
0.7500	1.0000	0.2500
0.8125	1.0000	0.1875
0.8750	1.0000	0.1250
0.9375	1.0000	0.0625
1.0000	1.0000	0
1.0000	0.9375	0
1.0000	0.8750	0
1.0000	0.8125	0
1.0000	0.7500	0
1.0000	0.6875	0
1.0000	0.6250	0
1.0000	0.5625	0
1.0000	0.5000	0
1.0000	0.4375	0
1.0000	0.3750	0
1.0000	0.3125	0
1.0000	0.2500	0
1.0000	0.1875	0
1.0000	0.1250	0
1.0000	0.0625	0
1.0000	0	0
0.9375	0	0
0.8750	0	0

0.8125	*0*	*0*
0.7500	*0*	*0*
0.6875	*0*	*0*
0.6250	*0*	*0*
0.5625	*0*	*0*
0.5000	*0*	*0*

We can see that the jet colormap assigns the lowest values to the blue, the highest to red and a combination of red, green and blue for intermediate values. Therefore, we can use any mathematical function to modify these maps. Two useful modifications to 'colormaps' can be obtained using powers or square roots. Look at the effect of these functions:

```
figure

imagesc(fluorescent_cells(:,:,1))
colorbar
colormap(hot.^2)

figure
imagesc(fluorescent_cells(:,:,1))
colorbar
colormap(sqrt(hot))
```

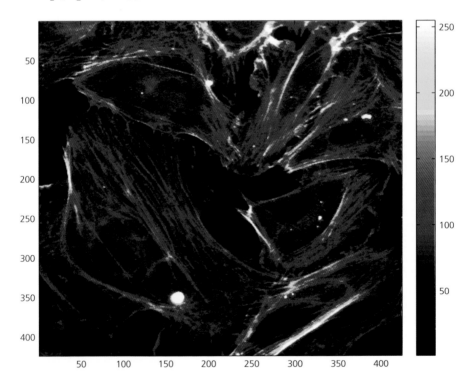

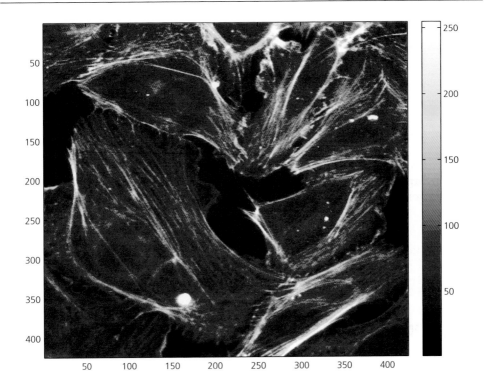

The values that form the colormaps are always between 0 and 1, therefore raising those values to the power of 2 makes them smaller (i.e. $0.5^2 = 0.25$) except for one as $1^2 = 1$, and thus the image looks darker. The square root makes the numbers bigger (i.e. square root of $0.5 = 0.7071$), again with the exception of 1, and the image looks brighter.

2.5   Thresholding and Manipulating Values of Images

As we explained in the introduction to Matlab, it is possible to use functions, either Matlab functions or others defined by the users, within functions such as 'imagesc'. For instance, if we compare the grey-scale image of the blood cells against a certain threshold we will generate a binary image in the following way:

```
figure

imagesc(blood_cells>160)
```

The previous command indicated to Matlab that it should compare every pixel of the variable 'blood_cells' against a value of 160. This will give a binary answer of 1 if the pixel value is higher and 0 if it is lower. The response is then displayed as a binary image where some parts of the cells are shown in white and others in black.

It is easy to modify the threshold level, and in some cases, such as fluorescent images, this process called "thresholding" can provide useful results.

Matlab can display colour images in several ways. The simplest way is to display a matrix in which the colour is stored as a combination of the three primary colours red, green and blue. To analyse the way that colour is formed in Matlab, we will again read two images from Wikipedia. The first image (uploaded to Wikimedia Commons by 'NicolasGrandjean', February 2006) contains one purple-coloured lymphocyte among other beige stained cells and a lighter beige background, and the second (uploaded to Wikimedia Commons by 'Lingulidas', March 2006) shows a few green-stained macrophages against a very light grey-green background:

```
white_cell_1 = imread(...
'http://upload.wikimedia.org/wikipedia/commons/1/17/Lymphocyte2.jpg');
white_cell_2 = imread(...
'http://upload.wikimedia.org/wikipedia/commons/1/15/Macrophage.jpg');
```

To display each of the images in a separate figure we simply use the command 'figure' before calling 'imagesc'.

```
figure
imagesc(white_cell_1)

figure
imagesc(white_cell_2)
```

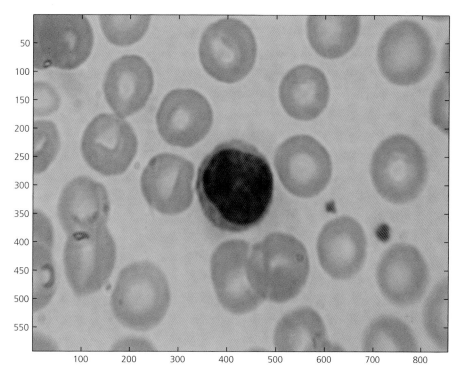

Nicolas Grandjean, February 2006; http://en.wikipedia.org/wiki/File:Lymphocyte2.jpg. Reproduced under a Creative Commons License.

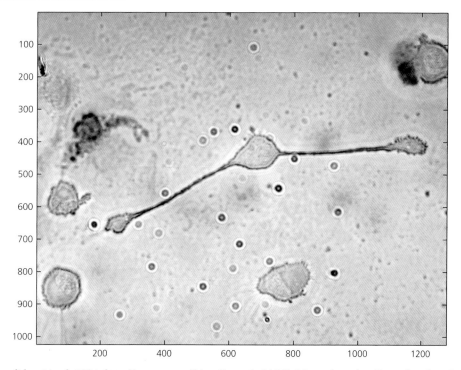

Lingulidas, March 2006; http://commons.wikimedia.org/wiki/File:Macrophage.jpg. Reproduced under a Creative Commons License.

There are several specialised tools to analyse images, which will be explored in later chapters. For the time being, we will observe the components of the channels by typing the coordinates of one pixel of each image into the Command Window. To observe the background we can select the pixel at row 400 and column 300 of the lymphocyte, but we do not select any particular level to observe the values in each of them:

```
white_cell_1(400,300,:)

    ans(:,:,1) =

       223

    ans(:,:,2) =

       193

    ans(:,:,3) =

       141
```

and the row 800 and column 300 of the macrophage

```
white_cell_2(800,300,:)

    ans(:,:,1) =

       229

    ans(:,:,2) =

       227

    ans(:,:,3) =

       206
```

The background of the lymphocyte (223, 193, 141) indicates a darker shade than the macrophage (229, 227, 206), as the second values are closer to (255, 255, 255), which corresponds to pure white. We can also notice that the lymphocyte has higher values in the first level, which describes the red component or red channel of the image, which is reflected in the beige tonality.

At this moment, we may return to analyse the components of the blood cells we had previously read from Wikipedia. That image only shows greys, no colours, however, it had three colour channels. To investigate further we can observe the values for the three channels of one pixel in the same way as before:

```
blood_cells(800,800,:)

    ans(:,:,1) =

       244
```

```
ans(:,:,2) =

    244

ans(:,:,3) =

    244
```

We can see that the three channels have identical values; this is the way in which grey is formed in the RGB colour space with equal contribution into the red, green and blue channels.

We can now compare the foreground of the strongly stained cells. We have seen that if we type the matrix of interest in the Command Window and press enter without a semicolon in the end of the line, Matlab will 'echo' the value of the matrix to the Command Window. We can also display values of more than one variable or matrix if we place them inside another matrix, and again not use the semicolon at the end, like this:

```
[white_cell_1(257,400,:)  white_cell_2(613,338,:)]


    ans(:,:,1) =

    140    0


    ans(:,:,2) =

    42    38

    ans(:,:,3) =

    99     0
```

The values of each channel of the two variables are displayed next to each other, and it is very easy to compare them. The lymphocyte has a high level of red, lower level of blue and lowest of green (140,42,99) which corresponds to the pink-purple hue, whilst the macrophage at that point has a pure level of green (0,38,0) and the low value (with respect to 255) indicates that it is a dark green. Another way of observing the colour channels is by displaying the channels separately:

```
figure
subplot(1,3,1)
imagesc(white_cell_1(:,:,1))
subplot(1,3,2)
imagesc(white_cell_1(:,:,2))
subplot(1,3,3)
imagesc(white_cell_1(:,:,3))
colormap(gray)
```

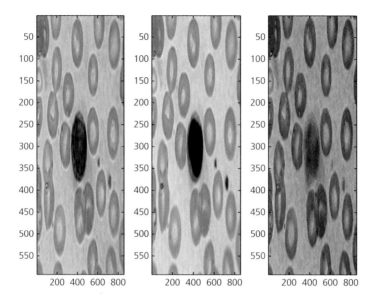

```
figure
subplot(1,3,1)
imagesc(white_cell_1(:,:,1))
colorbar
subplot(1,3,2)
imagesc(white_cell_1(:,:,2))
colorbar
subplot(1,3,3)
imagesc(white_cell_1(:,:,3))
colormap(gray)
colorbar
```

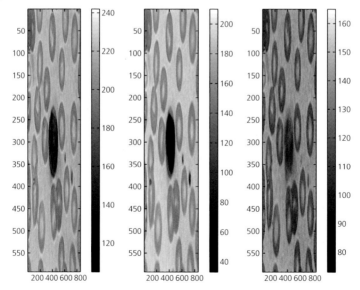

The colour channels of this image show how each pixel is formed by channels of different intensities, and each cell or region of a fairly homogeneous colour can be identified in different ways. In the present example, the leukocyte staining creates a high contrast in the blue channel as the pink-purple staining has a low component of blue compared with the rest of the image. A simple segmentation technique to separate this cell from the background and the other cells could be based on the green component, for instance:

```
segmented_cell_1                    = (white_cell_1(:,:,2)<90);
```

To superimpose the segmentation on the original image, we can subtract the segmentation from one of the channels. To do this, we will copy the channels one by one to a different variable and modify the green channel only to change the shading of the selected area.

```
figure
subplot(1,2,1)
imagesc(segmented_cell_1)
colormap(gray)

subplot(1,2,2)
segmented_cell_over(:,:,1)=white_cell_1(:,:,1);
segmented_cell_over(:,:,3)=white_cell_1(:,:,3);
segmented_cell_over(:,:,2)=white_cell_1(:,:,2)...
              +90*uint8(segmented_cell_1);
imagesc(segmented_cell_over)
```

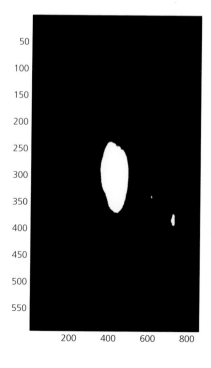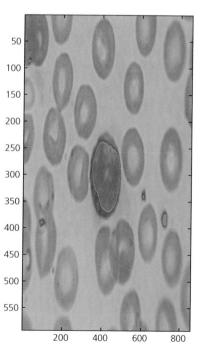

Notice that we had to convert the result of the segmentation, which is a 'logical' variable, into 'uint8', which is the same class of the original image. Here we displayed the logical image with black corresponding to 0 and white corresponding to white.

One of the problems of thresholding as a segmentation technique is the selection of the most suitable threshold. In the previous example, the selection of 90 as a threshold value included only the bright pink pixels. However, there are pixels that surround these, which could be considered as part of those cells. If we select instead 130 we could obtain a very different segmentation.

```
figure

segmented_cell_2            = (white_cell_1(:,:,2)<130);
segmented_cell_over(:,:,1)=white_cell_1(:,:,1);
segmented_cell_over(:,:,3)=white_cell_1(:,:,3);
segmented_cell_over(:,:,2)=white_cell_1(:,:,2)...
              +60*uint8(1-segmented_cell_2);

subplot(1,2,1)
imagesc(segmented_cell_2)
colormap(gray)
subplot(1,2,2)
imagesc(segmented_cell_over)
```

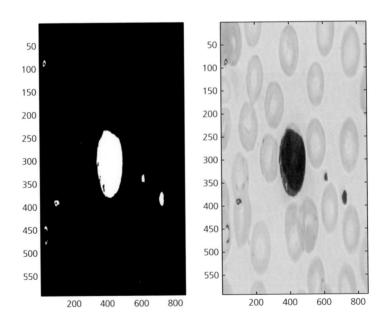

Notice that we have played with the way in which the results are overlaid on the original image. First the background maintained its original colour and the segmented region was modified. Then, the background was modified and the segmented cells have maintained their original colour.

To compare the two segmentations we can display them in the same figure by adding them together:

```
figure
imagesc(segmented_cell_1+segmented_cell_2)
colormap(gray)
```

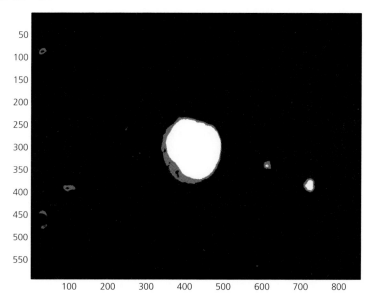

When we add the two results together, we will have a maximum value, displayed as white in the figure, where the two thresholds selected that region, grey where only one of them selected the region and black otherwise. Alternatively, we can subtract one result from the other like this:

```
figure
imagesc(segmented_cell_1-segmented_cell_2)
colormap(gray)
```

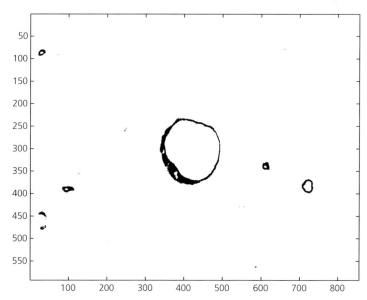

In the case of the subtraction, we are highlighting the regions that are different in the two variables with black, whilst the regions that are equal are displayed in white. This is a good way to compare results visually as the contrast between black and white emphasises the differences.

2.6 Converting Images into Doubles

This section will finish with an important concept. Images cannot be easily manipulated when their type is 'uint8', which is the default type for images. To perform operations with them it is better to convert images into 'double' type. This can be easily done either by changing an existing variable, for instance:

```
% Converts image to a matrix of doubles
blood_cells  = double(blood_cells);
```

or it can be done at the moment when the image is originally read like 'double(imread(...));'.

Once the images are stored as a double data type, they are treated and handled as matrices of doubles, so that any operations, such as addition, subtraction, multiplication, inversion, transposition, etc. are valid, as long as they follow matrix operations. For instance, it is not possible to use the function 'hist' to obtain histograms of uint8 types, but it is possible with doubles:

```
figure
hist(blood_cells(:),50);
axis tight
```

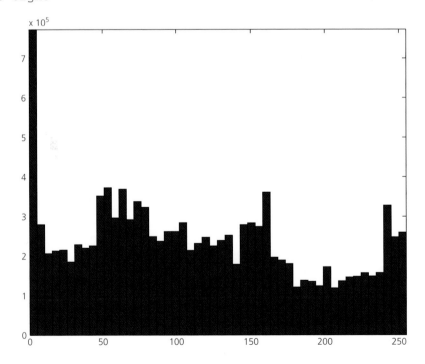

2.7   Save Your Code and Data

It is good programming practice to save your data and code frequently to avoid the possibility of losing any of them. The code written in the 'm-file' Editor can be saved by typing Control-S (Command-S in Macs) or using the menu FILE > SAVE and selecting a suitable file name (only for the first time). The data that you generate with the commands of Matlab can also be saved in different ways. You can export your figures as images as was explained in the previous section with the command 'print', which was stored in the shortcuts. You can specify the format (Tiff, BMP, JPG, etc.), the resolution in dots per inch and the name of the desired image, for example:

```
print('-dtiff','-r400','myFirstFigure.tif')
```

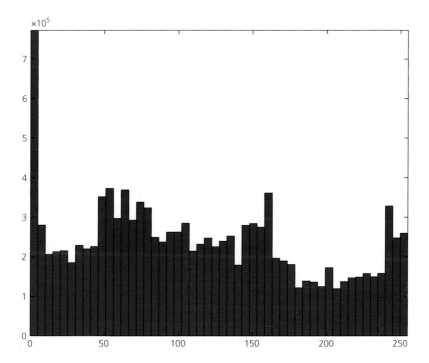

However, in some cases you may be interested in saving the data itself, that is, all the Matlab matrices that have been used to generate your data. Matlab allows this to be done with the command 'save'. If you simply type:

```
save
```

Matlab will save all the variables that are currently in the workspace into a file called 'Matlab.mat', which will be placed in the current working folder. You can also specify the name of the file you want to save and which, if not all, of the current variables, for example:

```
save('Author_CellLine_ExperimentNumber_Time_Date',...
   'blood_cells','fluorescent_cells')
```

We have specified the name of the file to be 'Author_CellLine_Experiment-Number_Date', and the only variables we saved in that file were blood_cells and fluorescent_cells. Notice that both name of file and names of variables have to be written between single quotes. In many cases, it is important to have a consistent definition of names for experiments, in case they form files that will later be used for analysis. For instance, if the names are to describe the times of a longitudinal study, these should all have a consistent unit, either seconds, minutes, hours, etc., e.g. Exp1_00min.mat, Exp1_10min.mat, Exp1_20min.mat, Exp1_30min.mat, etc. These names can later be used to process all the files in a single process or "batch". Batch processing will be described in a separate recipe in Chapter 7.

To retrieve your saved data you use the command 'load' and the name of the file. To demonstrate this process, we may want to first delete all the data from the current workspace with the command 'clear':

```
clear
```

The workspace is clear, but we can use the commands 'who' and 'whos' to display the current data from the Command Window:

```
who
```

Since the workspace is empty 'who' does not display any variables. Now we can load the data we previously saved, and try again:

```
load Author_CellLine_ExperimentNumber_Time_Date
```

```
who
```

```
Your variables are:
 blood_cells        fluorescent_cells
```

```
whos
```

Name	Size	Bytes	Class	Attributes
blood_cells	2239×1800×3	96724800	double	
fluorescent_cells	424×424×3	539328	uint8	

The command 'who' will list the names of the current variables, whilst 'whos' list the variables and displays the name, size, bytes and class (or data type) of each variable.

CHAPTER 3
Introduction to Colour

3.1 Mixing and Displaying Colours

Humans naturally perceive colour through the specialised cells of the retina. Cameras and other acquisition devices require special techniques to "perceive" colour such as optical filters and separate sensors for primary colours. Equivalently, display equipment uses different techniques to present colour. A common display technique relies on the mixing, that is adding or subtracting, of a reduced set of primary of fundamental colours from which other colours can be obtained. An initial requirement is thus an agreement of the primary colours to be used, and if these are going to be added or subtracted, this is normally called a "colour space" or "colour model" in which the primary colours and the combination rules are defined. The additive colour space uses red, green and blue (RGB) as primary colours, and mixes them in an additive way, in the way that light of different colours is added, that is, white becomes a mixture of colours, and black is the absence of colours. The subtractive colour space uses cyan, magenta and yellow (CMY) and in some cases, black (CMYK) as well, and mixes them in the way that inks are mixed. In the CMY colour model, the absence of the colours will produce white, and a combination of all primaries becomes black.

In this section we will mix and display colours in the RGB space. To do this we need a matrix with three levels so that the image that is displayed will have a triplet of values at each pixel.

To begin, let's add only a red component with different intensities to an image, or a two-dimensional (2D) plane. One way of doing this is to create a one-dimensional (1D) matrix of values that increase, and then multiply that with a matrix containing only ones. If we multiply these two 1D matrices in the correct order (see the introduction to Matlab for product of matrices), we can create the 2D plane with red values increasing horizontally:

```
figure
planeRed                    = zeros(11, 11, 3);
lineOfRed                   = 0:0.1:1;
planeRed(:, :, 1)           = ones(11, 1)*lineOfRed;
imagesc(planeRed)
```

Biomedical Image Analysis Recipes in MATLAB®: For Life Scientists and Engineers, First Edition.
Constantino Carlos Reyes-Aldasoro.
© 2015 John Wiley & Sons, Ltd. Published 2015 by John Wiley & Sons, Ltd.

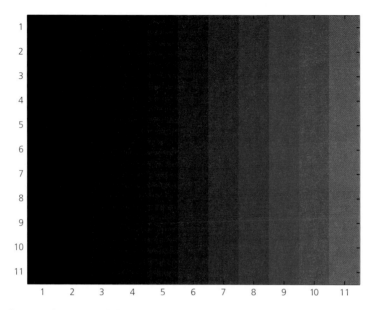

Notice that we have used the commands 'zeros' and 'ones' to generate matrices of given dimensions whose elements are all either 0 or 1, respectively.

In the same way that we created the scale of red intensity on the horizontal axis, we can create it in the vertical axis by transposing (turning a row into a column and vice versa) the matrix with a single apostrophe (') at the end of the matrix:

```
figure
planeRedGreen                = zeros(11, 11, 3);
lineOfGreen                  = 0:0.1:1;
planeRedGreen(:, :, 1)       = ones(11, 1)*lineOfRed;
planeRedGreen(:, :, 2)       = (ones(11, 1)*lineOfGreen)';
imagesc(planeRedGreen)
```

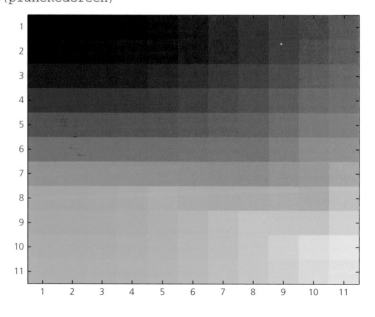

Since the matrix with the increase on the vertical values was placed on the second level, the change of intensity becomes a green variation, and when the two elements start to combine, the colours tend towards yellow.

To create blue shades, the 2D matrix should be placed in the third level of the three-dimensional (3D) matrix:

```
figure
planeRedBlue            = zeros(11, 11, 3);
lineOfBlue              = 0:0.1:1;
planeRedBlue(:, :, 1)   = ones(11, 1)*lineOfRed;
planeRedBlue(:, :, 3)   = (ones(11, 1)*lineOfBlue)';
imagesc(planeRedBlue)
```

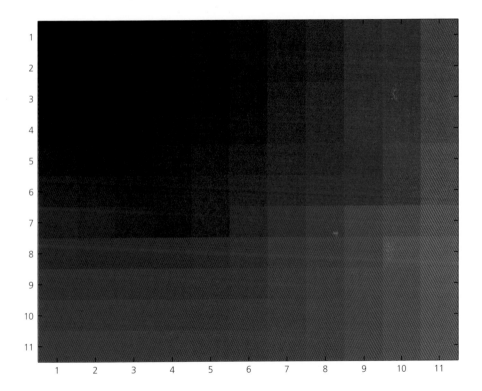

In the previous example the combination of red and blue will give pink shades. Finally we can combine green and blue in the same way:

```
figure
planeGreenBlue            = zeros(11, 11, 3);
planeGreenBlue(:, :, 2)   = (ones(11, 1)*lineOfGreen)';
planeGreenBlue(:, :, 3)   = (ones(11, 1)*lineOfBlue);
imagesc(planeGreenBlue)
```

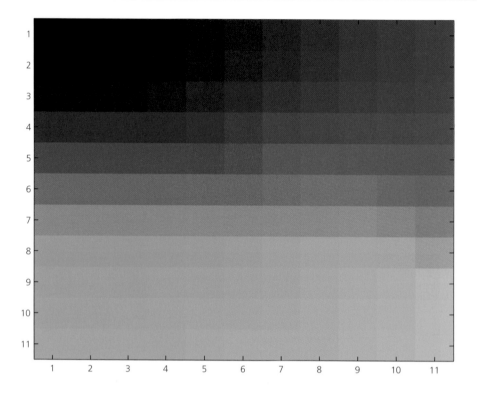

So far we have created 2D planes to display the combination of pairs of primary colour gradients, but since there are three of these, the combination of three values is better displayed in three dimensions. There are many ways to do this, but we will show how to do this in two particular ways. First, we can combine the planes generated so far and display them in 3D. To do this, we require coordinates in the 'x', 'y' and 'z' axes and the use of the command 'surf' to display the planes as surfaces. First, to generate the coordinates, there is a useful command called 'meshgrid', and we need to specify the dimension of the coordinates to use in the following way:

```
[xAxis, yAxis]                    = meshgrid(1:11, 1:11);
```

We still require values for the 'z' axis, and initially we want to display the base plane, that is, with the same vertical coordinates, so we create a matrix filled with values of 1:

```
zAxis                             = ones(11);
```

With the axes now generated, we can use 'surf' to display the first plane:

```
figure
surf(xAxis, yAxis, zAxis, planeRedGreen)
```

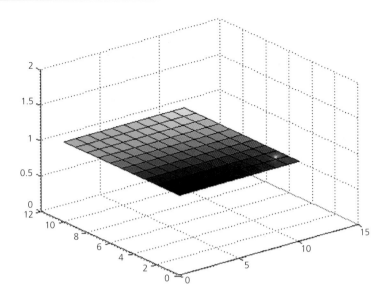

To add the two other planes in the same figure, we need to tell the figure to keep whatever had been added previously. We do that with the command 'hold on', and whenever we want to release the figure we use 'hold off', for instance:

```
hold on

surf(zAxis, yAxis, xAxis, planeGreenBlue)
surf(xAxis, zAxis, yAxis, planeRedBlue)

rotate3d on
view(130, 30)
```

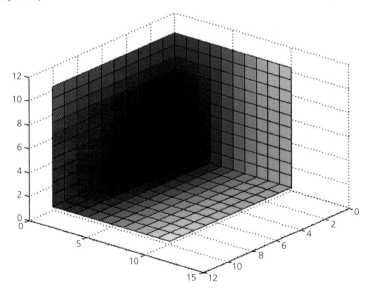

The figure now contains a 3D representation of the mixture of red (increasing towards the left), green (increasing towards the right) and blue (increasing vertically). So far we have only combined channels in pairs, and in one corner we have always had a value of (0, 0, 0). However, to reach pure white we would need all channels to reach one. Let's create one more plane; in this case all the elements are given a blue component of one, that is, a plane at the top of the cube displayed previously:

```
planeRedGreenBlue              = zeros(11, 11, 3);
planeRedGreenBlue(:, :, 1)     = ones(11, 1)*lineOfRed;
planeRedGreenBlue(:, :, 2)     = (ones(11, 1)*lineOfGreen)';
planeRedGreenBlue(:, :, 3)     = (ones(11));

surf(xAxis, yAxis, 11*zAxis, planeRedGreenBlue)
```

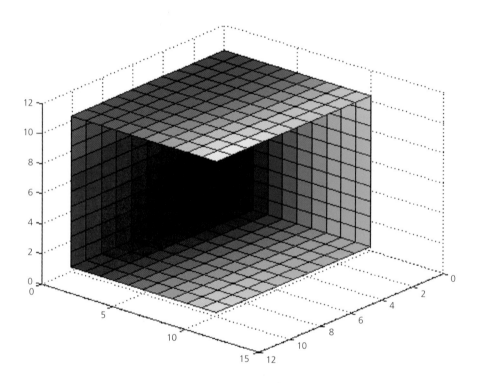

This example illustrates a common issue when dealing with 3D data sets, that is, the representation of some elements, may obstruct other elements that lie behind that element. Matlab allows the surfaces to have a controllable degree of transparency, and this is easily modified with the command 'alpha' where the parameter 1 is no transparency and 0 is total transparency. For example look at the image with transparency set to 0.5:

```
alpha (0.5)
```

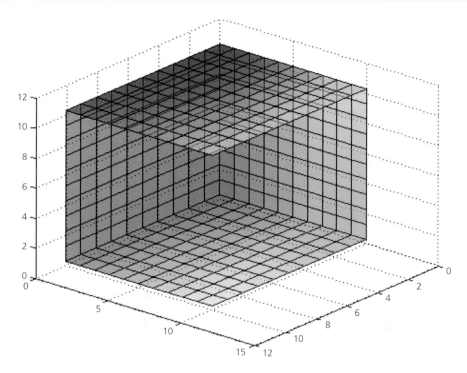

We can add more planes to form a cube, but we will now turn our attention to the second technique to display colours in three dimensions.

Another way to display colours in 3D is to place a single element, not a whole plane, at certain coordinates with certain colours. For instance, in a similar way to the previous cube with elements mixing red, green and blue, we could iterate over the three axes and place a sphere whose colour is linked to the position itself. For this, we can use the command 'scatter3', which takes as input the 'x', 'y' and 'z' positions as the first three input arguments, then the size (we use 200 for this example) and then the colour of the element, which we use the same value as its position. The final parameter specifies that the spheres should be displayed filled and not just in outline:

```
figure
hold on
for x=0.0:0.2:1
  for y=0:0.2:1
    for z=0:0.2:1
      scatter3(x, y, z, 200, [x, y, z], 'filled');
    end
  end
end


rotate3d on
view(160, 60)
```

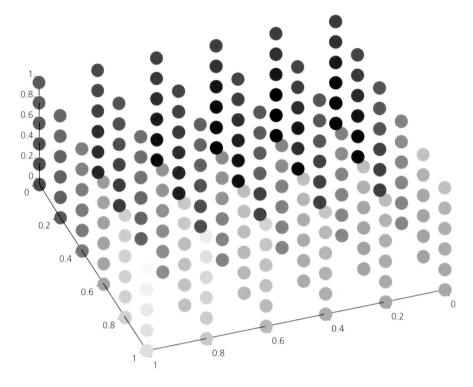

The combination of these display techniques can create very nice images of high quality. We will return to use these in future recipes.

CHAPTER 4

Western Blots

One of the most common techniques used in molecular biology for detection of DNA, RNA, proteins, their combinations and modifications is the use of gel electrophoresis, commonly known as "blots". The original blots were developed by Edwin Southern and thus became known as Southern Blots. Subsequent gel electrophoretic techniques kept the term blot and changed the "direction": Western, used to detect proteins; Northern, used to detect RNA or mRNA; Eastern, to detect protein post-translational modifications like lipids. The directional analogy has been maintained with newer techniques such as North-Western, Middle Eastern and Far-Western.

In Western blots, the macromolecules of interest will be separated by their size or charge. The molecules travel through a porous medium, the gel, and some will travel faster than others and may halt halfway through the migration process. Subsequently, an antibody specific to the protein of interest is allowed to bind to the protein of interest, and a secondary antibody system is utilised to provide a fluorescent signal proportional to the amount of target protein-antibody binding. The fluorescent signal was traditionally used to expose a sheet of film and the density of the blot measured by photodensitometry. Advances in imaging technology mean that today a digital CCD imager can be used to generate electronic data recordings of the resulting signal intensity without the need of films.

Western blots are very commonly used to assess the protein characteristics of disease or the molecular response to potential therapies. The differences in the molecules become apparent as the presence or absence of bands (in what sometimes is called one-dimensional electrophoresis) or spots in two-dimensional electrophoresis. The intensity and size of the bands is associated with the quantity of the molecules, and their location, relative to known locations, and is related to their size or charge. The signal intensity of the target protein can be normalised against the signal intensity of a (presumed) universally expressed known protein, which is typically actin.

Biomedical Image Analysis Recipes in MATLAB®: For Life Scientists and Engineers, First Edition.
Constantino Carlos Reyes-Aldasoro.
© 2015 John Wiley & Sons, Ltd. Published 2015 by John Wiley & Sons, Ltd.

4.1   Recipe 1: Many Ways to Display a Western Blot

A blot is commonly shown as a grey-scale image, which contains a series of dark regions on a bright background, or its negative, with bright regions on a dark background, for example:

```
WesternBlot_1 = imread('WesternBlot1.jpg');
WesternBlot_2 = imread('WesternBlot2.jpg');
```

```
figure
imagesc(WesternBlot_1)
colormap(gray)
```

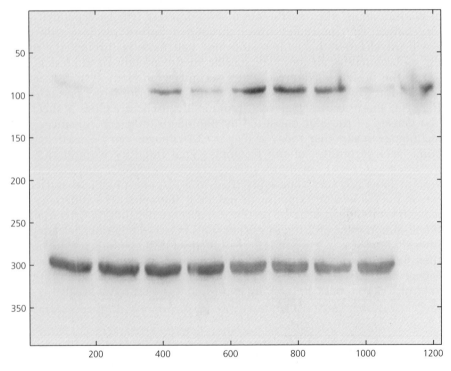

Image courtesy of Dr. Chryso Kanthou, The University of Sheffield, UK.

```
figure
imagesc(WesternBlot_2)
colormap(gray)
```

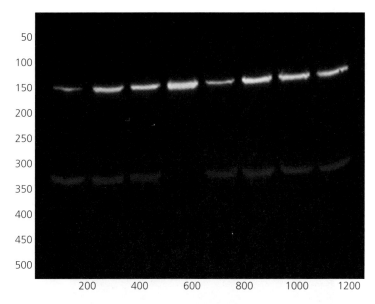

Image courtesy of Dr. Chryso Kanthou, The University of Sheffield, UK.

Notice the command 'colormap' after the display of the image with 'imagesc' was used to change the colormap to 'gray' in order to display the blot in grey scale as it is commonly done. In Matlab we have the flexibility to display images in many ways, which in some cases can help visualise any differences that are otherwise difficult to perceive. For example, let's change the display from grey scale to a display with a set of rainbow set of colours. This is called the 'jet' colourmap in Matlab:

```
colormap(jet)
colorbar
```

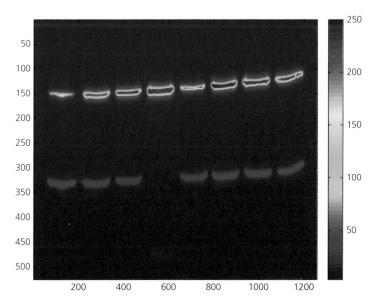

The blot is now displayed with dark blue as the background and warm colours for higher (brighter) intensities. It is important to notice that this is only changing the way the image is displayed; the data of the image itself remains unchanged. Try other colormaps that may be suitable in other context; for example, a common map is 'hot', which is similar to 'gray' in the sense that it shows dark as black, and bright as white, but instead of using shades of grey, it uses "warm" tones, e.g. red, orange, yellow, etc. We can obtain a 'hot' map like this:

```
colormap(hot)
```

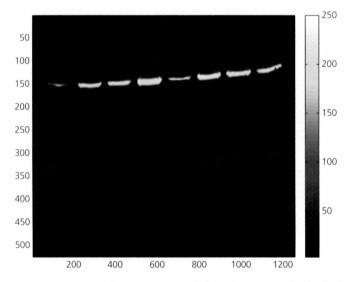

The colour maps 'gray' and 'hot' are useful as they provide high discrimination when the images are printed with black and white printers. The counterpart to 'hot' is the colour map that uses "cold" tones, e.g. cyan, purple, violet, etc. This is obtained with the map 'cool' like this:

```
colormap(cool)
```

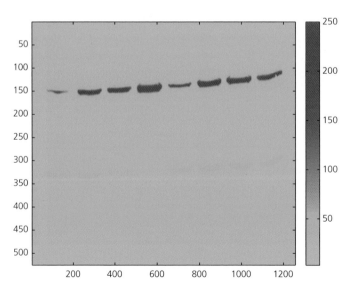

Whilst this colour map may not be as good as others in terms of the discrimination of the intensities on the screen, it may be useful for printing purposes as the dark background is displayed as a bright shade.

One final colour map will be presented, but it is recommended that you explore the other options like 'bone', 'autumn', 'spring', etc. The colour map 'hsv' has a wide range of hues to map the intensities of an image. This may be useful, as intensities that are very close to each other may correspond to visually distinctive hues. For instance, look how the intensities of the bands of the blot change with this colour map:

```
colormap(hsv)
```

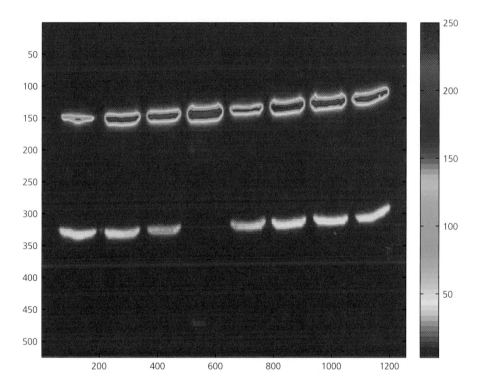

The colours of the colormap are predefined by Matlab but can easily be modified in several ways. One of the most convenient ways is by a user interface that shows the colours, which can be modified with a slider. The interface is generated with the command 'colormapeditor':

```
colormap(jet)
colormapeditor
```

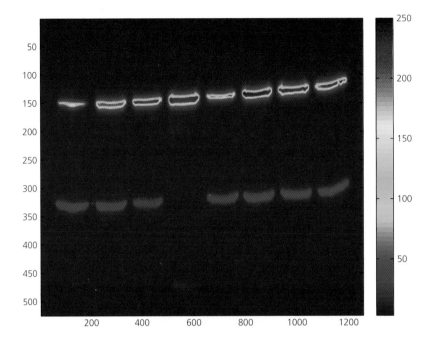

The window that appears with colormapeditor contains a slider graph that describes the colormap: the intensity values of the image are scaled and divided into 64 intervals, and each of these intervals is assigned a colour, which is then used to display the image on the figure. When you move the cursor over the graph, the values corresponding to the Index (1–64), intensity of the data CData (0–251 for this example) and the RGB (red-green-blue) and HSV (hue-saturation-value) are displayed on the "Current color info" box. Below the graph, there are several "anchor" points. These anchor points can be moved along the horizontal axis to modify the colormap. Try moving them to produce different results.

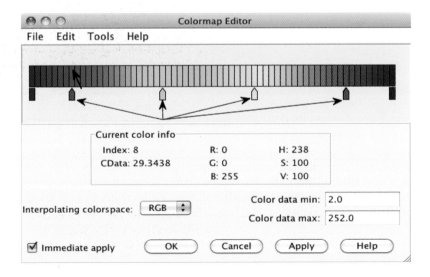

Another useful way to display a Blot image is to display it as a 3D "mesh", where the intensity of the pixels is translated to height and the height is displayed with the colour of the corresponding colour map. To do this, we use the command 'mesh'. There are two important considerations that we need to take into account before displaying the intensity as a mesh: (1) the dimensions of the data and (2) the "data type". The command 'mesh' assigns a height according to the intensity of each pixel; therefore we cannot pass as an input argument an image with more than one level, e.g. 3D matrices. Second, as the heights can be any value, and not just values between 0 and 255, which are the values normally used with images, the command 'mesh' requires that the data to be displayed be of data type 'double' instead of 'uint8' (un-signed integers of 8 bits). We can concatenate the conversion and the display in one single instruction line:

```
figure
mesh(double(WesternBlot_2))
```

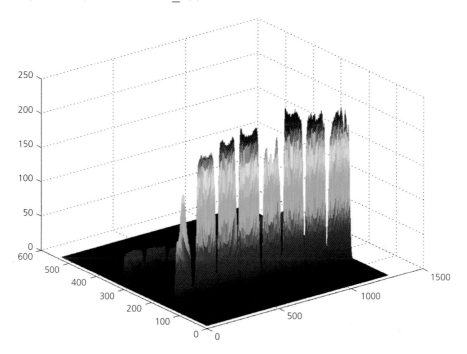

A careful observer may notice that there are some differences between the way the Blot has been displayed by mesh and the image. Whilst the faint band appeared on the bottom of the image, it appears on the "top" of the mesh, however, the thinner narrower band of the bright row is towards the left in both cases. The reason behind these differences is the way the axes have been used, as it was explained in Chapter 2 (2.3 Displaying Images). In addition, there are regions of the plot that are not covered by the data. We can change the current axes from 'xy' to 'ij' and then fit the data to the axes tightly with the following commands:

```
axis ij
axis tight
```

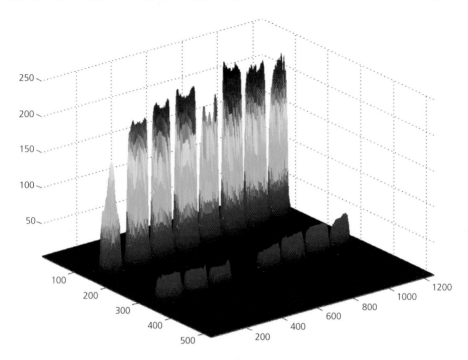

We can now change the point of view from which we are observing the plot. This rotation is obtained with the command 'view' followed by the angle and azimuth values like this:

```
view(10,60)
```

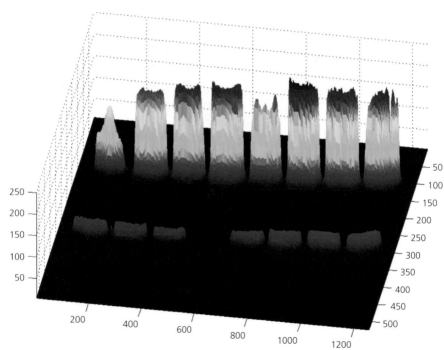

We can change the point of view manually as well. To do this we need to enable the rotation of the figure with the command 'rotate3d' and the argument 'on' in a single instruction line 'rotate3d on'. To disable the rotation we simply turn it off like this: 'rotate3d off'. You should plot your own figure at this moment and modify the values of view and rotate the figure with the mouse to see the differences while visualising the blot image.

As a final modification of the mesh, we can change the viewpoint to have it above the mesh:

```
view(0,90)
colormap(gray)
```

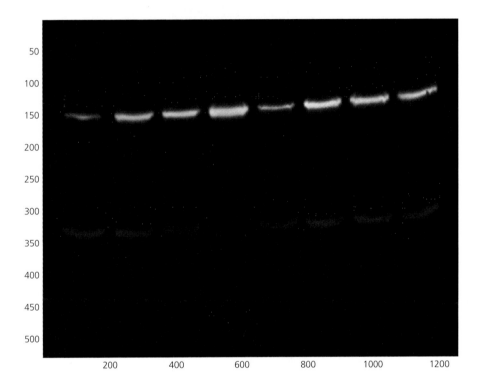

Since we also changed the colour map, we have returned to the same image as the one that was generated with the command 'imagesc' at the beginning of this recipe. However, the images are not exactly the same; if you repeat this process it may be the case that you will notice a few white dots scattered around the background and wonder where these came from, as the background should be all black. The reason that some white spots appear is due to the nature of the mesh and the resolution of the screen. We can illustrate this by zooming in to a smaller region of the image and displaying it again as a mesh from two points of view:

```
figure
subplot(121)
mesh(double(WesternBlot_2(130:160,590:630)))
axis ij
```

```
axis tight
subplot(122)
mesh(double(WesternBlot_2(130:160,590:630)))
axis ij
axis tight
view(0,90)
colormap(gray)
```

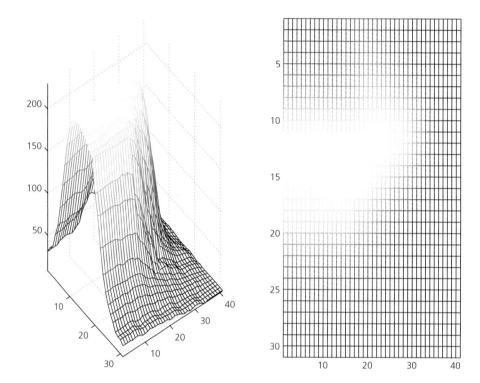

It should be clear now that the command 'mesh' is displaying a "mesh" in which the pixels are assigned a location, and these locations are connected with the "mesh", and there is nothing in between the lines of the mesh. If the previous figures seemed to be full, it was due to the compact nature of the mesh. If we want to have a value in between the lines of the mesh, we are creating a "surface", and we need to use the command 'surf' like this:

```
figure
surf(double(WesternBlot_2(130:160,590:630)))
axis ij
axis tight
```

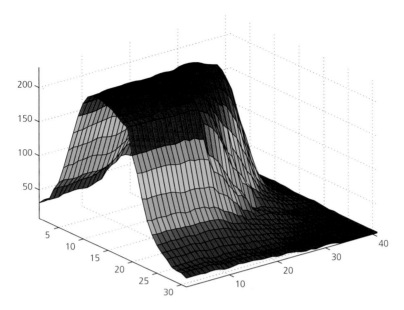

As we mentioned before, it is not possible to use 3D data with the commands 'mesh' and 'surf' as these commands will translate the intensity of each pixel into height. If a matrix has three levels, we need to select just one of them to be displayed, for example:

```
figure
mesh(double(WesternBlot_1(:,:,1)))
axis ij
view(10,60)
axis tight
```

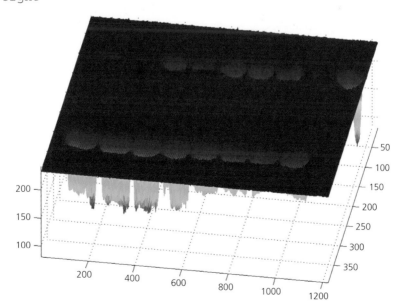

Since 'WesternBlot_1' has a bright background and dark bands, the background appears as a high plateau and the bands as holes on the "ground". This may not be the best way to visualise the data, so it may be more convenient to invert the intensity axis. This can be easily done in several ways. One of them is plotting the negative values of the data using the minus sign "−" when passing the image as an argument to the command 'mesh':

```
figure
mesh(-double(WesternBlot_1(:,:,1)))
axis ij
view(10,60)
axis tight
```

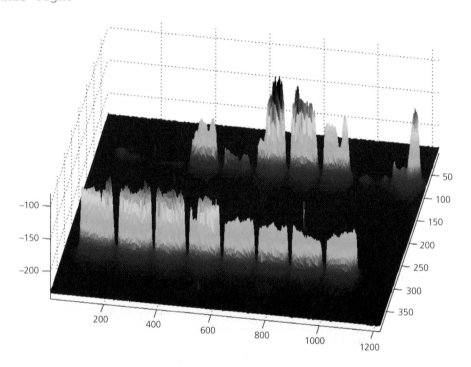

The figure now has the background at the bottom and the bands rising up, which is a better way to visualise the data. Alternatively, we can change the direction of the 'z-axis'. To do this, we need to capture the "handle" of the plot. In Matlab each "element" such as a figure, an image or an axis is defined by its "handle". Handles will be explored in more detail in Chapter 8, but we will introduce simple modifications with the use of handles here.

The handle contains a large amount of information that define the element or object itself. To obtain the handle of the current figure we use the command 'gcf' that stands for "get current figure", and to obtain the handle of an axis we use the command 'gca', which stands for "get current axis". We can modify some of the properties of an object by modifying the handles. For example, we will display the same figure as before and will modify the vertical axis so that it is displayed in reverse order:

```
figure
mesh(double(WesternBlot_1(:,:,1)))
axis ij
view(10,60)
axis tight
set(gca,'Zdir','Reverse')
```

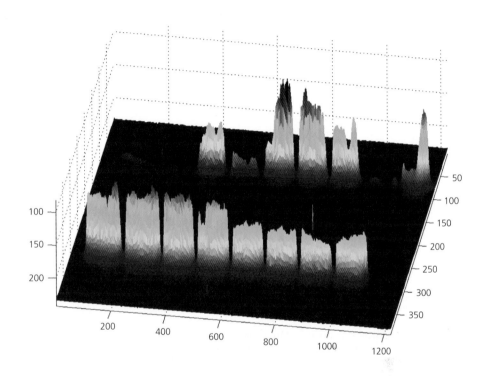

   Notice that this time, the colours of the original display have been maintained. The background is red and the peaks (or troughs) of the bands tend towards the blue, whilst when we changed the data the background was blue and the peaks tended towards the red.
   Now that we have a better understanding of the nature of the data, in terms of the intensity values of the image, a natural step is to try to visualise how the bands are distributed, and then answer questions such as how dark/bright each band is and how small/large they are.
   Another useful way to display images is to draw contours at different "heights" or regions of constant intensity. Contours are commonly used in topographical maps to display changes in altitude. Matlab can display these contours with the command 'contour' like this:

```
figure
contour(WesternBlot_1(:,:,1))
axis ij
```

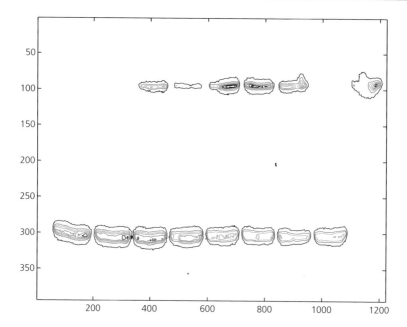

The previous figure shows six contours that were automatically determined by the maximum and minimum levels from the intensity range of the data. It is possible to determine the number of contours desired, or the positions of the contours, by passing a second parameter to the contour command. If it is a single number, it will correspond to the number of contours; if it is a matrix with many numbers, it corresponds to the heights at which the contours are desired. For instance, to generate 15 contours we would use the following command:

```
contour(WesternBlot_1(:,:,1),15)
axis ij
```

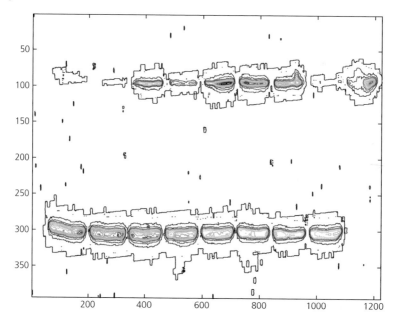

and to generate 3 contours at specific values we might use the following command:

```
contour(WesternBlot_1(:,:,1),[135 215 233])
axis ij
```

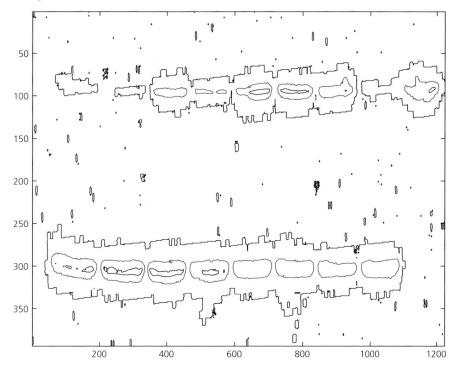

The previous images show several important characteristics of the images them-selves and also highlight some of the issues and difficulties of analysing images. One of the problems is selecting an adequate cut-off point (or threshold) for a particular analysis. We will investigate that in some later recipes in this and other chapters.

4.2   Recipe 2: Investigating the Numbers That Make a Western Blot

Recipe 1 showed many ways to display an image, a Western Blot in particular, but some times we are interested not only in displaying images but investigating some of their characteristics. To do this, the first step is to look into some numbers, those that define the size, the values, and so on. We will describe some of these techniques in this recipe.

We begin again by loading the data that we want to investigate:

```
WesternBlot_1 = imread('WesternBlot1.jpg');
WesternBlot_2 = imread('WesternBlot2.jpg');
```

We can first analyse the dimension of each image. We do this with the command 'size':

```
size(WesternBlot_1)
size(WesternBlot_2)
```

```
ans =

        394            1223               3

ans =

        525            1256
```

We can now see that 'WesternBlot_1' has 482,560 picture elements, commonly known as pixels, arranged in 394 rows, 1223 columns. We can also see that this image contains 3 levels, which correspond to the colour channels of the image, red, green and blue (more on colour later), which totals 1,445,586 values of intensity for this image.

'WesternBlot_2' on the other hand has 659,400 pixels arranged in 525 rows, 1296 columns and only 1 level. Why does one image have 3 levels while the other has only 1 level when they look very similar as grey-scale images? This has to do with the way the data is encoded to generate images with different image file formats like JPG, TIFF, BMP, etc. Images with colour will generally have three levels with different values for each level. Grey scale images can either have a single channel, which describes the grey-level data or three identical red, green and blue channels. It is important to clarify that a true black-and-white image will consist of pixels that are only black or white, without any other brightness (grey) levels in between.

To verify that the three channels of 'WesternBlot_1' have the same values, one simple way to investigate is to select a region, say the first five rows and columns. To display the values on the Command Window we simply type:

```
WesternBlot_1(1:5,1:5,:)
```

```
    ans(:,:,1) =

        234   232   231   232   233
        234   232   231   232   233
        234   232   231   232   233
        234   233   231   232   233
        234   233   232   232   233

    ans(:,:,2) =

        234   232   231   232   233
        234   232   231   232   233
        234   232   231   232   233
        234   233   231   232   233
        234   233   232   232   233

    ans(:,:,3) =

        234   232   231   232   233
        234   232   231   232   233
```

```
234  232  231  232  233
234  233  231  232  233
234  233  232  232  233
```

The values of the three levels are identical; that is why it is grey.

To investigate the actual values of the pixels of one of the images, we can display them in the Command Window. We can select a small region that corresponds to one of the bands rows (say 91 to 95) and columns (671 to 675). As we know that all the levels contain the same information, we only need to display one of them:

```
WesternBlot_1(91:95,671:675,1)
```

```
ans =

  178  178  179  180  181
  154  154  154  155  156
  124  124  126  129  128
   99  100  103  108  107
   85   86   91   97   97
```

Printing values of variables in the Command Window is simple and effective, but cumbersome if we want to investigate a whole image or do this repeatedly. There are two alternative options, which are more effective. First we can analyse the values of a variable in the "Variable Editor". To open a variable you can simply double-click on the variable name in the Workspace in the Desktop. Alternatively you can use the command 'openvar' in the following way:

```
openvar WesternBlot_2
```

The previous command opens a new window where all the values of the variable 'WesternBlot_2' are displayed in a matrix similar to a spreadsheet. It is possible to navigate through this matrix and to modify values there.

The second tool to analyse the values of the variable, as well as to obtain interesting information is to use the command 'imtool'. When we type 'imtool', it will create a new figure where the data is displayed and several point-and-click tools then become available:

imtool(WesternBlot_2)

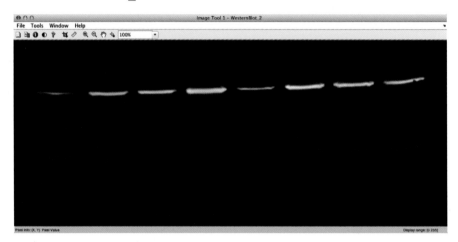

The command 'imtool' has several very useful tools available; some of these are displayed in the next figure and are labelled 1 to 5. To start with, when 'imtool' is launched, the image is displayed and the position and value of the pixel at the tip of the cursor is displayed at the bottom left of the figure. You can move the cursor around the image to investigate the values of the pixels.

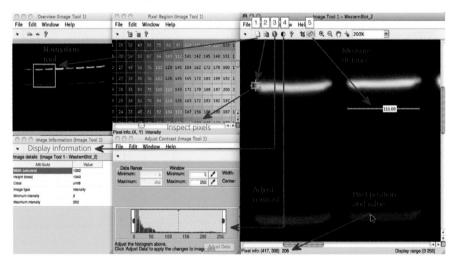

If the image is large and you want to zoom into a section of the image you can use the Navigation tool (1), which will open a new window with the whole image and a region of interest (ROI) that can be moved around to select a specific region of the image. The dimension of a box that defines the region of interest can be modified by selecting the icons on the toolbar: the left button zooms in, the right button zooms out.

To investigate the actual values of the pixels, the second icon on the toolbar will create a small ROI window on the image and will display the values of the pixels inside that ROI in a new window. The size and position of the ROI can be modified with the mouse.

The general characteristics of the image are displayed on a separate window with the information tool (3). The information displayed corresponds to the width, height, data type and intensity levels.

It is possible to modify the contrast of the image with icon (4) on the toolbar. This icon will open a new window where the histogram of the image is displayed. It is possible to modify the maximum, minimum and central value of the histogram to improve the contrast of the image displayed in 'imtool'.

The last tool with the label (5) is the distance tool. When this icon is clicked, it allows the user to overlay distance lines on the original image. The distance lines will indicate the distance between the end points and are useful when the manual measurement of a distance is required.

The interactive 'imtool' is very useful for manually exploring image data, so it is worthwhile spending some time experimenting with the many options available.

4.3   Recipe 3: Image Histograms

Histograms are one of the most common and useful tools used to interpret the data of images. The construction of a histogram is a technique inherited from statistics, where the range of intensity values is divided in a number of containers or "bins". The data is then assigned to the bins according to the intensity values, and the number of values assigned to each bin is counted. In this way, the distribution of the data, according to intensity values, is obtained. Histograms are normally displayed with the bins on the horizontal axis and the counts of values, the frequency, on the vertical axis.

Histograms are important as they can reveal something about the nature of the image. They have become very common and are sometimes displayed on the screens of digital cameras as a tool to improve the exposure of photographs. However, in this recipe we will see that they are very useful for analysing images.

Obtaining histograms in Matlab is very easy; we can begin by loading the Western Blots of the previous recipes plus a third one:

```
WesternBlot_1 = imread('WesternBlot1.jpg');
WesternBlot_2 = imread('WesternBlot2.jpg');
WesternBlot_3 = imread('WesternBlot3.jpg');
```

The command to obtain a histogram of the data is 'hist'. If we pass an image as an argument to 'hist' it will obtain the histogram of every column of the image, which is not what we are after in most of the cases. To obtain the histogram of the whole image we can easily convert the data from a 2D or 3D matrix into 1D (one row) by typing "(:)" after the name of the variable, for example:

```
A = [1 2 3 4; 5 6 7 8; 9 10 11 12];
A(:)
```

```
ans =

     1
     5
     9
     2
     6
    10
     3
     7
    11
     4
     8
    12
```

Notice the order in which the rearrangement was done. It starts at the first row and column, and then goes down through the first column, then the second, etc. Before we can use the 'hist' command, there is another important consideration. The command 'hist', as many other Matlab commands that perform mathematical operations, requires that the data type used is of type 'double', and since images are normally stored as integers, of type 'uint8' (un-signed integers of 8 bits) we need to convert them first with the command 'double'.

The following set of commands will display the three Western Blots and their corresponding histograms under them in a single figure.

```
figure
subplot(2,3,1)
imagesc(WesternBlot_1)

subplot(2,3,2)
imagesc(WesternBlot_2)

subplot(2,3,3)
imagesc(WesternBlot_3)

subplot(2,3,4)
hist(double(WesternBlot_1(:)),(1:10:255))
axis tight; grid on

subplot(2,3,5)
hist(double(WesternBlot_2(:)),(1:10:255))
axis tight; grid on

subplot(2,3,6)
hist(double(WesternBlot_3(:)),(1:10:255))
axis tight; grid on

colormap(gray)
```

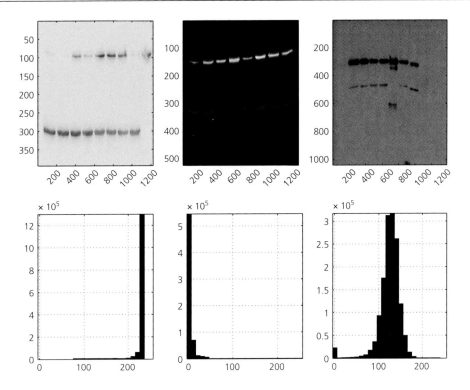

Image courtesy of Dr. Chryso Kanthou, The University of Sheffield, UK.

The figure shows several important characteristics of the images. The background of the first blot is bright, which corresponds to a majority of the pixels being assigned to the bins on the right side of the histogram. The second blot shows the opposite behaviour; it has a dark background and the histogram peak on the left-hand side.

The third blot is different again and the majority of its pixels are assigned to intermediate bins and a minority appeared on the left of the histogram. If we look at the image of the blot, we can see that the bands are dark and the background is very noisy. The majority of the image is formed of a noisy background of medium grey shades; that is why there is a bell-shaped curve in the centre of the histogram, indicating a range of intensity values around an average. This is a typical characteristic of Gaussian image noise.

The previous example obtained the histograms of the three example images with the following arguments: first, the image of the Western Blot itself, and second, a 1D matrix specified by a range syntax: (1:10:255). This second argument defines the way in which the histogram is calculated and controls the way the "bins" of the histogram are created. If we do not pass a second argument, 'hist' will calculate by default a histogram with 10 bins, equally spaced between the minimum and maximum values of the data:

```
figure
hist(double(WesternBlot_3(:)))
```

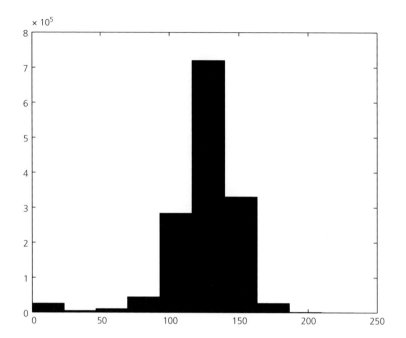

If we want a higher or lower number of bins, we can specify the number of bins by passing a single number, for example, 50:

```
hist(double(WesternBlot_3(:)),50)
```

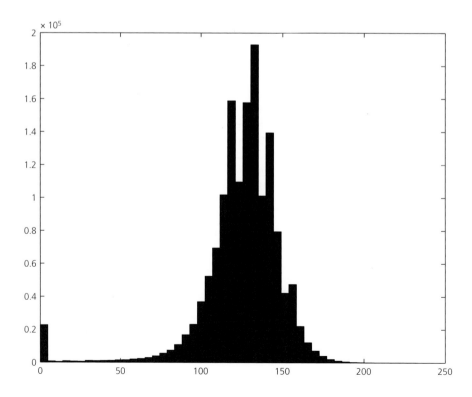

If we want to distribute the bins beyond the minimum or maximum range of the input data, or we want a specific distribution, we can indicate the centres of the bins with a 1D matrix such as [1 2 ... 255], where there is one bin for each possible grey level of an image, or its compact way as (1:255). If we want to reduce the number of bins, but still define the extreme values, we can use something like (1:5:255), which specifies a start bin at 1 and an end bin at 255 in steps of 5.

We may wonder why we need to specify the range for the bins. One case might be when we want to compare two images and one of them is brighter than the other. In that case, we specify the same bins for both images so that we can compare them with precisely the same ranges. Another case may be when we do not want to have a uniform distribution of the bins; we may know that the data is concentrated in a certain region and are interested in "zooming" in on the histogram bins in the range of intensities without increasing the total number of bins.

Now let's try to extract more information from one of the histograms. In the previous commands we only displayed the histograms, that is the distribution of the bins on the horizontal axis and the number of pixels per bin in the vertical axis. We can assign those values to new variables when using hist in the following way:

```
[yWestern,xWestern] = hist(double(WesternBlot_1(:)),5:5:250);
```

We have now stored the histogram values in the variables 'xWestern', 'yWestern'. Since we already know that the image is mostly bright and the histogram has a very high peak in the high values, we might want to modify the way we display the histogram, we will see one technique that provides a better visual analysis of the distribution of the intensities. We do this by "damping" the large values of the histogram. This can be done with the mathematical function of a logarithm. For the following figure we will use the command 'plot' to display the values of 'xWestern' on the horizontal axis against 'yWestern' on the vertical axis.

```
figure

subplot(211)
% Define the line as a blue (b) solid line (−) with
%    round markers (o)
plot(xWestern,yWestern,'b-o')
% Add a grid and tighten the axis
grid on
axis tight
title('Linear scale of the vertical axis')

subplot(212)
plot(xWestern,yWestern,'b-o')
% Change the scale on the vertical axis to logarithmic
set(gca,'yscale','log')
% Add a grid and tighten the axis
grid on
axis tight
title('Logarithmic scale of the vertical axis')
```

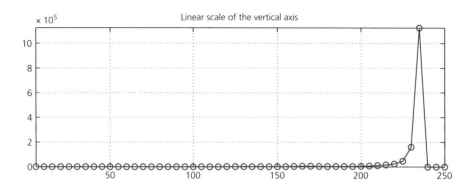

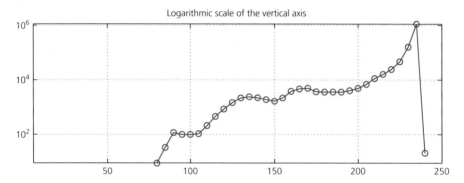

Notice that to transform the plot, we used again the command 'gca' and then we changed 'yscale' to 'log'.

The previous figure shows how different a plot with a logarithmic scale can look from one with a linear scale. Whilst the linear plot seems to have lost all the detail for values lower than 200 of intensity, in the logarithmic plot we can see how these values are distributed. Notice that there are no points for the extreme regions. The histogram has detected no values in these regions, and therefore the variable 'yWestern' has a series of zeros, the logarithm of zero corresponds to minus infinity, so Matlab will not display these points.

The normal way to display a graph is with a linear scale, in which the intervals are all equally spaced. The logarithmic scale will not be linear, but rather will increase with orders of magnitude; for our previous example, the "ticks" on the vertical axis are [10, 100, 1000, 10,000 up to 10,000,000], which are written in scientific notation, e.g. [10^1, 10^2, etc.].

If we want to analyse the values of the histogram, we can modify the "ticks" of the horizontal scale, again by setting the "handles" of the current plot like this:

```
set(gca,'xtick',(20:20:250))
```

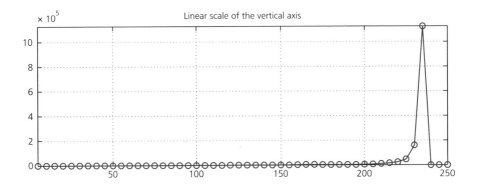

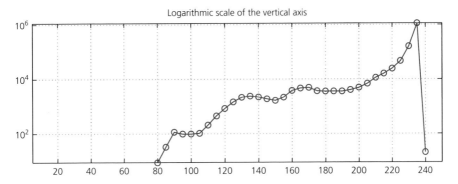

Since we are using a logarithmic scale on the vertical axis, we can see with more detail how the data of the Western Blot is distributed. The bin with the largest number of pixels is the one with intensity 235 with slightly more than a million elements. That is, nearly 80% of the image will have pixel intensities close to 235. It is worth noting that the histogram was counting all the elements of the image, so for each pixel we will have three identical values that were stored in the red, green and blue channels.

The counts of the histogram drop quickly on both sides. The bins centred at 230 and 240 counted 161,349 and 21 elements. Towards the centre of the histogram we can see three "peaks" or "humps" (technically known as "local maxima") at 90, 135 and 170 and three valleys (local minima) at 100, 150 and 185. Another important value is around 220, where the slope changes its inclination, which suggests that the pixels with values above 220 will be part of the background and those below that correspond to the bands of the blot.

These values can help us explore the levels at which we want to observe contours (see Recipe 2) because, as we have seen, selecting the correct threshold level is important. For instance we can try 100, 150 and 220:

```
figure
contour(WesternBlot_1(:,:,1),[100 150 220])
axis ij
```

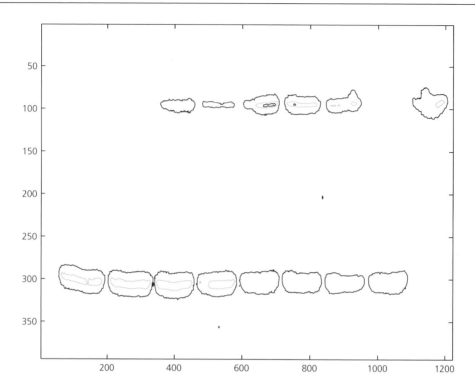

The contours selected with the threshold of 100 will only include the very dark regions of the Western Blot, and these are very small. The contours selected with 150 are now distributed within the darker bands and are absent in lighter bands, and 220 has successfully separated the bands from the background. This could be a starting point for a quantitative analysis.

4.4   Recipe 4: Transforming an Image of a Western Blot

In the previous recipes we have displayed the images in different ways and we have investigated their values. However, we have not yet modified the data. This recipe will introduce some of the simplest transformations like changing the values of certain pixels of the image. The complexity and usefulness of these transformations will increase gradually with further recipes.

A transformation of the image refers to a method in which an original image, say 'A', will be modified into a new image, 'B'. The modification will either change the values of the image (and leave the dimensions unchanged), change the dimensions of the image (and leave the values unchanged) or change both dimensions and values. Cropping an image will discard some of the pixels and create a new image with smaller dimensions. A comparison of the values of an image against a fixed value or threshold will create a new image where the output image will describe whether the pixels were above or below the threshold.

The process of comparing against a threshold is called "thresholding" and is a very popular analysis technique due to its simplicity, and it can provide interesting results. Thresholding can be either based on global (all the image) or local information, and more than one threshold can be used. The philosophy behind thresholding is that pixels with a grey level below or above a chosen threshold belong to one region and the remaining pixels to other region. In any case the idea is to partition into regions, object/background, or object a/object b/.../background. Thresholding methods rely on the assumption that the objects to separate are distinctive in their intensity levels and often use the histogram information, thus ignoring spatial arrangement of pixels. Although in many cases good results can be obtained, in some particular cases the intensities of certain structures are often not uniform, and simple thresholding can divide a single structure into different regions. Another matter to consider is the noise intrinsic to the images that can lead to a misclassification. In many cases the optimal selection of an optimal threshold is a nontrivial matter.

Let's now apply thresholding to a Western Blot. We will use the same image as in the previous recipes and will use some of the values obtained in Recipe 3 for the thresholding.

```
WesternBlot_1 = imread('WesternBlot1.jpg');
WesternBlot_2 = imread('WesternBlot2.jpg');

WesternBlot_A = (WesternBlot_1(:,:,1)<220);

figure
imagesc(WesternBlot_A)
colormap(gray)
```

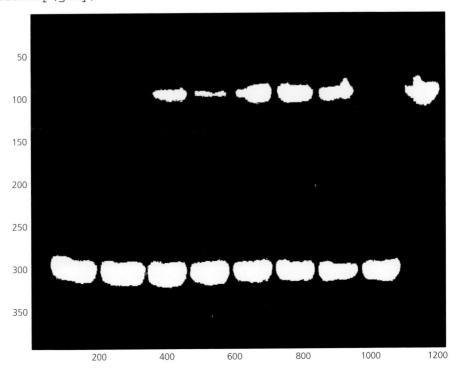

The previous lines compared pixels of the first level of the Western Blot against a threshold of 220. We only used the first level, as we know that the three levels have identical values since this is a grey-scale image. The image displayed shows a series of bands, defined by the pixels with values lower than 220. The shape of the bands is the same as one of the contours generated in recipe 3.

Besides changing the level of the threshold, there are other transformations that might be interesting. We might be interested in selecting only some regions of intensity and removing others. We can easily do this by addressing the values of the matrices (read "Addressing a matrix" in Chapter 1 for a description). For instance, we might be interested in looking at the bands described by the previous threshold but with the intensities and not only the contours. We use the following commands to do that:

```
WesternBlot_B = WesternBlot_1(:,:,1);
WesternBlot_B(WesternBlot_B>220)=0;

figure
imagesc(WesternBlot_B)
colormap(gray)
```

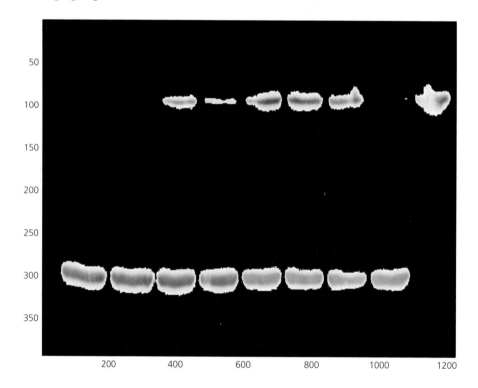

We have now set to zero all those pixels whose intensity values was above 220, the background, but have left unchanged the rest of the image. Since we are visualising the bands it may be evident that there are a few scattered pixels that do not form part of the bands but have values above 220, which may suggest that we could lower the threshold to remove these noisy pixels.

We can also do the opposite, remove those pixels below a threshold and leave the ones above unchanged:

```
WesternBlot_C = WesternBlot_1(:,:,1);
WesternBlot_C(WesternBlot_C<220)=0;

figure
imagesc(WesternBlot_C)
colormap(gray)
```

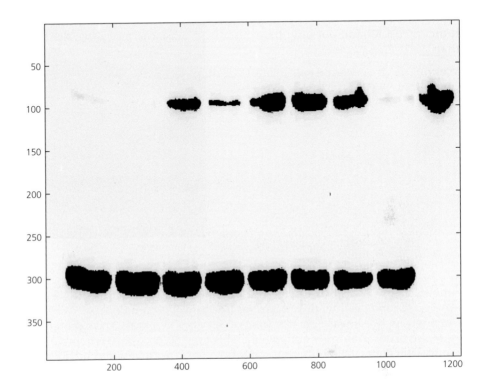

The previous transformations can be combined to remove pixels below one threshold and above another one or to remove pixels in the range between the two thresholds.

A different transformation can be obtained by modifying all the pixels with some common rule, the simplest would be to add or subtract a value to/from the image. A simple transformation that can be obtained by adding or subtracting is to obtain the negative of an image or to invert the intensity values. This is useful when we want to compare blots that have alternate dark or bright backgrounds as those of our previous examples. To invert one of our blots we can type:

```
WesternBlot_1_neg = 255-WesternBlot_1;

figure
subplot(121)
imagesc(WesternBlot_1)
subplot(122)
imagesc(WesternBlot_1_neg)
```

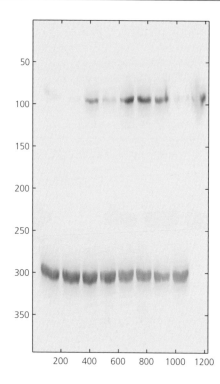 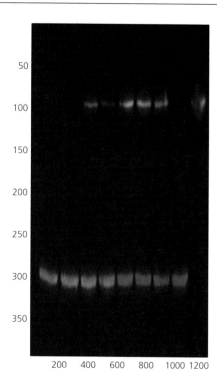

Although these transformation are useful, addition and subtraction are not as interesting as multiplication, division, roots and powers.

When we apply a linear operator (one of $+,-,*,/$) to a matrix, we modify all the values of the pixels uniformly and evenly. This may be necessary when we want to normalise data; it may be in different ranges. A classic normalisation technique used in statistics is to divide a population of values, like an image, by its mean value, which creates a new matrix with a mean value of zero. Then, that matrix is divided by its standard deviation, to create another matrix with zero mean and standard deviation of one.

In our case, we can use non-linear operators (like powers or roots) to change the values of the pixels so that they create an effect of brightening or darkening an image, for example:

```
WesternBlot_D = double(WesternBlot_1(:,:,1));
WesternBlot_E = sqrt(WesternBlot_D);
WesternBlot_F =      (WesternBlot_D.^2);
```

These lines have created three new matrices, one with the first level of Western Blot 1. We transformed the data into doubles as we will perform many operations that cannot be done with the original data type of the image ('uint8'). Then we obtained the square root of all the elements and stored that in a new matrix and squared the values as well. Notice that one command will act on every single value of the image. Traditional programming languages like C or Java would require loops that would scan the image pixel-by-pixel to do any such transformation. Matlab is far simpler to use, as it needs just one command for the whole image.

We now need to normalise the three matrices, as their ranges of intensities are very different. The following lines will normalise them so that the values are within 0 and 255.

```
WesternBlot_D = 255*WesternBlot_D/max(WesternBlot_D(:));
WesternBlot_E = 255*WesternBlot_E/max(WesternBlot_E(:));
WesternBlot_F = 255*WesternBlot_F/max(WesternBlot_F(:));
```

To visualise the effects of the transformations, we can display them as a single image. We can concatenate them very easily by placing them side-to-side in a single new matrix like this:

```
figure
imagesc([WesternBlot_D WesternBlot_E WesternBlot_F] )
colormap(gray)
```

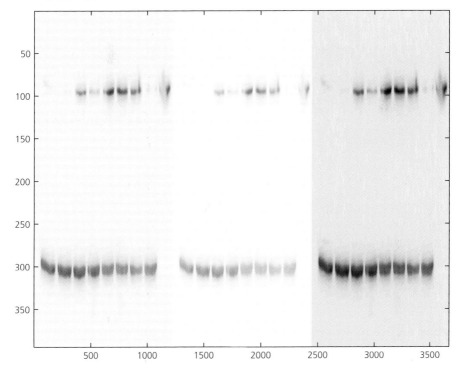

We can see that 'WesternBlot_E' (in the centre) is brighter than the original image (on the left) and 'WesternBlotF' (on the right) is darker. We can also observe the effect of transformations on the image histograms. We will first calculate the histograms of the three images and store them in separate variables.

```
[yWesternD,xWesternD] = hist((WesternBlot_D(:)),(1:10:255));
[yWesternE,xWesternE] = hist((WesternBlot_E(:)),(1:10:255));
[yWesternF,xWesternF] = hist((WesternBlot_F(:)),(1:10:255));
```

Now we can display the three histograms with the command 'plot', where we will pass three arguments of each line in the following order: horizontal values, vertical values, line specifications. To make the three lines distinctive, we will use different colours (blue, red, black), markers (circle, "x" and diamond) and line styles (solid, dash-dot, dashed).

```
figure
plot(xWesternD,yWesternD,'b-o',...
    xWesternE,yWesternE,'r-.x',...
    xWesternF,yWesternF,'k--d')
```

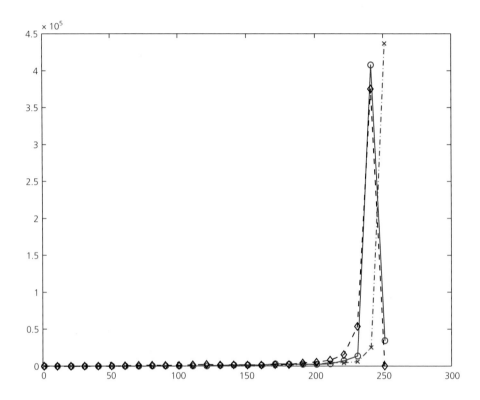

As we did before, we will change the scale on the vertical axis to logarithmic and also change the number of ticks on the horizontal axis to have a better comparison of the three lines. Notice that we have broken the instruction line into three lines by using three consecutive dots ('...') that indicates to Matlab that the instruction continues in the next line.

We will add a legend to the figure to clarify which line corresponds to which image.

```
set(gca,'yscale','log')
set(gca,'xtick',(20:20:250))
grid on
axis tight
legend('Original','WesternBlot E','WesternBlot F',...
    'location','NW')
```

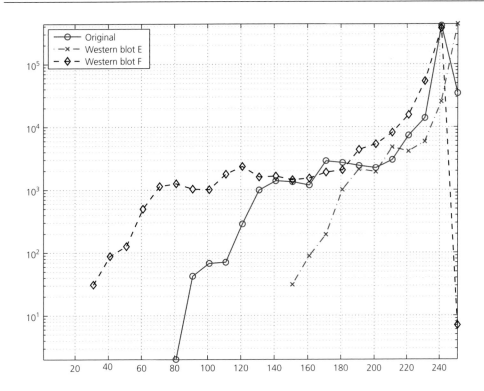

It is clear now how the square root shifted the histogram of image 'WesternBlot_E' towards the brighter regions (right), while the squaring operation shifted the histogram of 'WesternBlot_F' towards the darker regions (left).

4.5   Recipe 5: Quantification of the Data

In the previous recipes we have learned how to display images in different ways, to investigate the values that form our images, to transform them into new matrices with different characteristics and to use histograms. In this recipe we will learn how to extract information from our data.

We will begin with the thresholded version of the Western Blot as in recipe 4:

```
WesternBlot_1 = imread('WesternBlot1.jpg');

WesternBlot_A = double(WesternBlot_1(:,:,1));
WesternBlot_B = (WesternBlot_A<220);

figure
imagesc(WesternBlot_B)
colormap(gray)
```

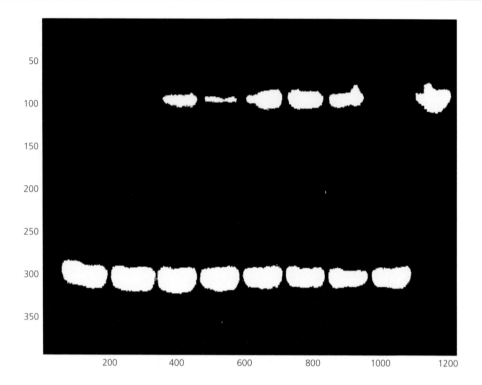

The thresholding has separated a series of bands from the background, and there are many algorithms that are useful to quantify our results so far. One of the first measurements that we might be interested is counting the number of bands, or of regions that were produced by the thresholding. To do this, we can first distinguish all the areas of connected pixels and assign them a label to distinguish them. This process is called labelling and obtained with the command 'bwlabel' in the following way:

```
[WesternBlot_labelled, numBands]      = bwlabel(WesternBlot_B);
```

We can now display the labelled image and display also the number of bands in the title of the image. The title takes as argument a 'string' variable (a series of characters) so to include numbers in a string we need to convert them from numbers to strings. We can do this with the command 'num2str' (number-to-string) and then we will concatenate the text and the numbers with the command 'strcat', which stands for "string concatenation".

```
figure
imagesc(WesternBlot_labelled)
title(strcat('This Western Blot has ', ...
    num2str(numBands),' Bands'))
```

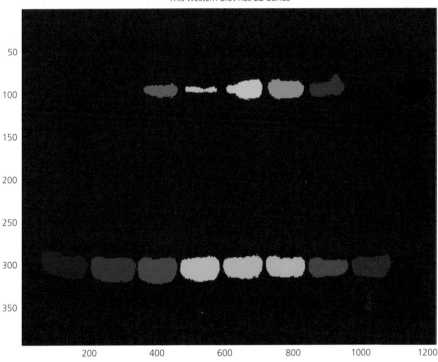

This western blot has 22 bands

Our figure now displays an image with 22 objects, one for every band, and each band has a different colour (or label). A quick manual count reveals that there are only 14 larger bands, therefore the thresholding process was not perfect and created some very small objects that do not correspond to any band of the Western Blot.

One easy way to improve the results would be to modify the threshold, and thus try to match the visual count of the bands. However, there may be more efficient ways to deal with the segmentation once we have generated separate objects from the bands (a more difficult problem is to deal with bands that are merged). If we observe carefully, we can see some tiny objects that are clearly not bands of the blot.

In order to identify these small regions (outliers), we can calculate the area of each of the objects previously generated. The command 'regionprops' calculates properties of the regions, and one of many options is to count their areas. We do this in the following way:

```
WesternBlot_props    = regionprops(WesternBlot_labelled,...
                            'Area');
```

Besides the area, 'regionprops' can determine many other important measurements and properties of the objects of a labelled image: perimeter, centroid, minor and major axis, eccentricity or Euler number just to name a few. The output of 'regionprops' is saved into a new matrix of a special data type called a 'struct' (short for structure) in which each property that was calculated (like the area) can be accessed by using a dot ('.') in between the name of the structure and the name of the property, e.g. 'WesternBlot_props.Area', 'WesternBlot_props.Perimeter', etc. To see the values of the areas we can place the structure in between square brackets and pass that as an argument to the command 'disp' like this:

```
disp([WesternBlot_props.Area])
```

```
  Columns 1 through 6

      3396        3450           2        3099        1202           1

  Columns 7 through 12

      2878         497           4        2611           1        1759

  Columns 13 through 18

         3           3        2462        1942           6        2177

  Columns 19 through 22

      1662        2511           1        2310
```

So now it is clear that while the objects that correspond to bands have areas of thousands of pixels, there are some regions with fewer than 10. Our next step naturally is to discard those small areas. This can be done by using two commands and our previous results. First we need to identify which objects we want to remove, those with less than a certain area, say 10 for our example, and link them to the label that they have been assigned. To do this we must use the command 'find':

```
find([WesternBlot_props.Area]>10)
```

```
  ans =

    Columns 1 through 13

      1    2    4    5    7    8   10   12   15   16   18   19   20

    Column 14

     22
```

We now know which are the regions that have been identified as having more than 10 pixels. Then, we use the command 'ismember' to select only a subset of all the labelled objects; the subset will be those objects that we determined previously with the command 'find'. Thus, we will store the new objects in a separate variable like this:

```
WesternBlot_ObjsRemoved = ismember(WesternBlot_labelled,...
    find([WesternBlot_props.Area]>10));
```

Now we label the new variable from which the small objects have been removed and we count again:

```
[WesternBlot_labelled, numBands]    = ...
  bwlabel(WesternBlot_ObjsRemoved);
```

```
figure
imagesc(WesternBlot_labelled)
title(strcat('This Western Blot has ',num2str(numBands), ...
  'Bands'))
```

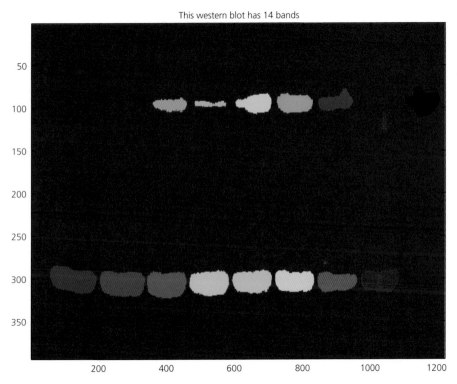

When a satisfactory segmentation of the bands has been obtained, we can now investigate some properties of each band. Since we want the same properties for each band, we can run a "loop". In Matlab one way to run a loop is to place a series of commands that will be repeated in between one line that starts with the command 'for' and another line with the command 'end'. The line with the 'for' will define an initial and a final value, and it may define the steps between initial and final values, for instance, to loop between 2 and 20 every four values, a 'for' cycle would be defined as:

```
for counter = 2:4:20
    % write some code here
    % this code will be repeated a number of times
```

```
    % and use the variable 'counter',
    % which will take the values 2, 6, 10, 14 and 18
end
```

For our example, we want to iterate over all the bands so we will use a variable called 'counterBands', which will take the value of each band one at a time. The results of our measurements will be stored in a matrix called 'BandCharacteristics'. We will now present the code with comment lines intercalated to explain each instruction line.

```
for counterBands = 1:numBands
    % Select one band from the labelled image
    currentBand              =   (WesternBlot_labelled== ...
                                 counterBands);
    % Extract those pixels of the image that correspond to the
    % current band defined by the counter, to do this we
    %   address
    % the matrix with the matrix 'currentBand'
    currentBandValues        =   WesternBlot_A(currentBand);
    % Now that we have the pixels of the band, we can
    %   calculate
    % a series of measurements, we start with the average
    % value of the pixels. We store this in the first
    % column of the output matrix
    BandCharacteristics (counterBands,1) = ...
        mean(currentBandValues);
    % Calculate the standard deviation of the
    % intensity and store in second column
    BandCharacteristics (counterBands,2) = ...
        std(currentBandValues);
    % Calculate the area as the number of pixels of the object
    % use the command 'numel' (number of elements) and
    % store in the third column
    BandCharacteristics (counterBands,3) = ...
        numel(currentBandValues);
    % find the centroid of each band and store in columns 4,5
    [currentRows,currentColumns]              = find(currentBand);
    BandCharacteristics (counterBands,4) = ...
        round(mean(currentRows));
    BandCharacteristics (counterBands,5) = ...
        round(mean(currentColumns));
end
```

With the measurements that we have obtained, we can start to infer more on the nature of our data. For instance we might compare the size of the bands with the intensity. To do this we can plot one of the measurements against the other:

```
figure
plot(BandCharacteristics(:,3),...
```

```
     BandCharacteristics(:,1),'r*','markersize',9)
grid on
axis tight
xlabel('Area of the Band','fontsize',16)
ylabel('Average Intensity of the Band','fontsize',16)
```

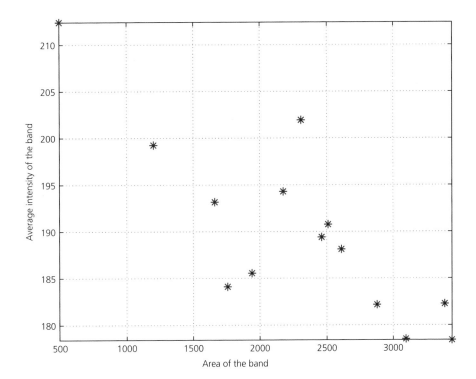

There appears to be an inverse correlation between size and intensity as the larger bands have lower intensities. Furthermore, one of the bands, the one with the smallest area (497) and highest intensity (212) is relatively distant to the others, which may correspond to an outlier or the fact that it is a band that is not really present. We should remember that this segmentation was obtained with a threshold value of 220. If we changed that value, this band may disappear. We can look at the standard deviation against the intensity to further probe this theory.

```
figure
plot(BandCharacteristics(:,2),...
     BandCharacteristics(:,1),'r*','markersize',9)
grid on
axis tight
xlabel('Standard deviation of the intensity','fontsize',16)
ylabel('Average Intensity of the Band','fontsize',16)
```

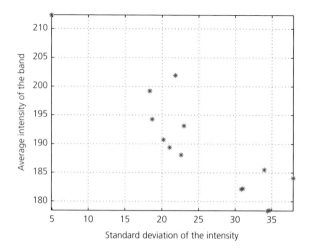

We can see that unlike the other bands where the standard deviation is larger, this band has a very small deviation, which again indicates that this band may be an outlier and could be removed from the selected bands.

When we compare the band intensity with its vertical position we can discover also an interesting trend, which shows that there is a clear difference between the size of the bands in each region.

```
figure
plot(BandCharacteristics(:,3),...
    BandCharacteristics(:,4),'r*','markersize',9)
grid on

xlabel('Area of the Band','fontsize',16)
ylabel('Vertical centroid of the band','fontsize',16)
```

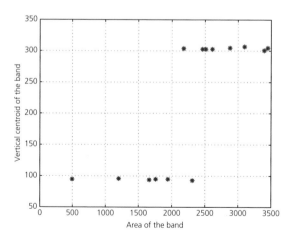

As a final analysis plot, we can overlay some of the measurements that we have just calculated over the original image. To do this, we will find the boundaries of the segmented bands with the command 'bwperim' like this:

```
WesternBlot_perimeters   = bwperim(WesternBlot_labelled>0);
WesternBlot_perimeters   = imdilate(WesternBlot_perimeters, ...
[1 1;1 1]);
```

Notice that the argument that was passed to 'bwperim' was compared against zero, which will set all the labelled bands to one. We have also dilated the perimeters so that they are easier to visualise later on.

We can now copy the original image into a new variable and we will modify only one channel of the new image so that the boundaries appear with colour over the grey-scale image. Since the original image is formed of pixels with the 'uint8' data type, we need to convert the boundaries to 'uint8' as well.

```
WesternBlot_1_segmented           = WesternBlot_1;
WesternBlot_1_segmented(:,:,2)   = ...
  WesternBlot_1_segmented(:,:,2).* ...
   uint8(1-WesternBlot_perimeters);

figure
imagesc(WesternBlot_1_segmented)
```

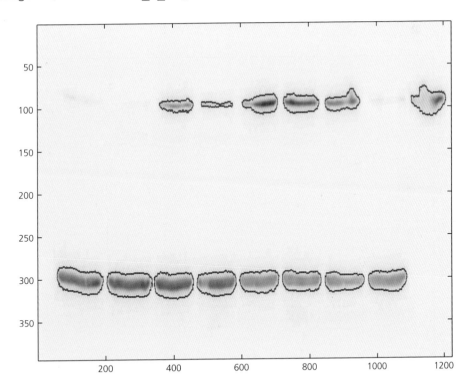

To conclude this Western Blot analysis we can display the information that we obtained previously, the average intensity and the area of the bands on the same image. To do this we need to use the command 'text' and the image we previously obtained. Again we will loop over the bands and will use the positions that we had previously calculated to place the texts. We will shift them up and left of the centroids by 50 pixels to avoid overlapping with the band itself.

```
for counterBands = 1:numBands
    % extract the position of the current band
    % from the matrix BandCharacteristics
    currentRow      = BandCharacteristics(counterBands,4);
    currentCol      = BandCharacteristics(counterBands,5);
    % extract the area and intensity of the
    % current band from the matrix BandCharacteristics
    areaText        = strcat('A:',num2str(...
        round(BandCharacteristics(counterBands,3))));
    intensityText   = strcat('I:',num2str(...
        round(BandCharacteristics(counterBands,1))));
    % display the information on the image
    % with the segmented bands, offset the position
    % so that the text is above the bands
    text(currentCol-50,currentRow-50,areaText,'fontsize',9);
    text(currentCol-50,currentRow-30,intensityText, ...
        'fontsize',9);
end
```

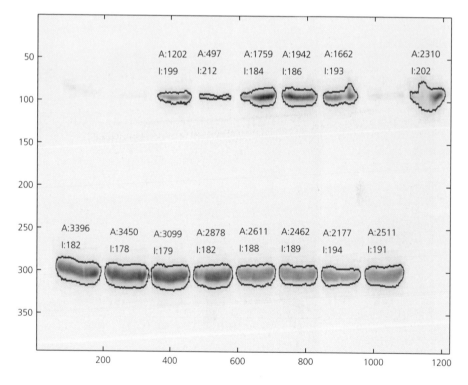

Notice how the text was formed by concatenating strings with the command 'strcat' and transforming the measurements into strings with the command 'num2str'. We can now observe carefully the band we had previously considered as an outlier. This is the second from the left on the top row. It can be seen that it is very narrow and does not correspond to a band that can be easily visualised. This can confirm the previous analysis and we could discard it in a subsequent analysis.

4.6   Recipe 6: Investigating Position of Bands

A visual examination of some blots can reveal the absence of intensity in a certain band; in some cases it is interesting to compare a certain band with one experiment and the same band with a different experiment. In this recipe we will look at how to determine the positions of bands with the help of some simple mathematical operations.

We begin this recipe by loading one image of a Western Blot and displaying it in grey scale:

```
WesternBlot_2 = imread('WesternBlot2.jpg');


figure
imagesc(WesternBlot_2)
colormap(gray)
```

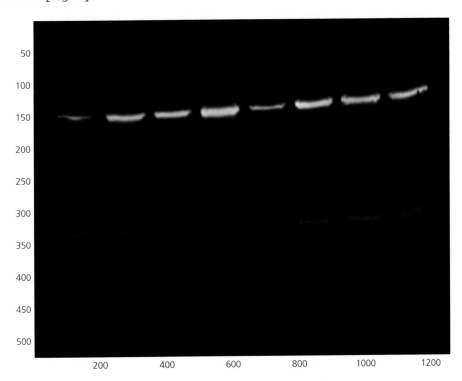

A quick visual examination reveals that the bands in the upper part of the image have higher intensities than those in the lower part. Also, we can see that not all bands have the same intensities. This can be better visualised if we change the colormap to one that has a better discrimination for human vision:

```
colormap(gray.^0.4)
colorbar
```

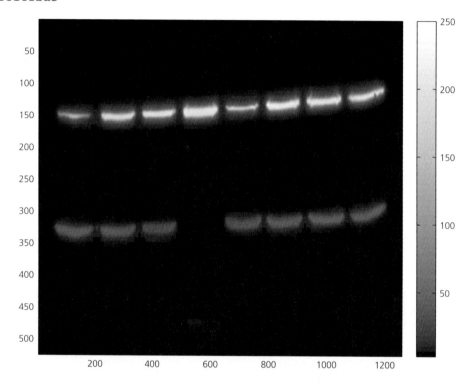

With this colormap, it becomes apparent that the fourth band from left to right has a very intense band in the upper row and the corresponding fourth band on the lower part is totally absent.

To locate the position of the bands, we can scan every column of the image to find the maximum value of that column:

```
% find the maximum in the first dimension
WesternBlot_maxColumn   = max(WesternBlot_2,[],1);
```

The command 'max' finds the maximum value along a certain specified dimension, in this case down the rows for every column. In the same way we can find the minimum with 'min', the average with 'mean', the sum with 'sum' and the standard deviation with 'std'. Let's now display the result. Notice that since this was an operation on all the columns, the data is no longer in 2D but a single 1D matrix, sometimes called "vector" or "array" and thus we use plot to display it:

```
figure                           % open a new figure
plot(WesternBlot_maxColumn)      % display a variable in 1D
grid on                          % add a grid to the display
axis tight                       % remove space of the plot
```

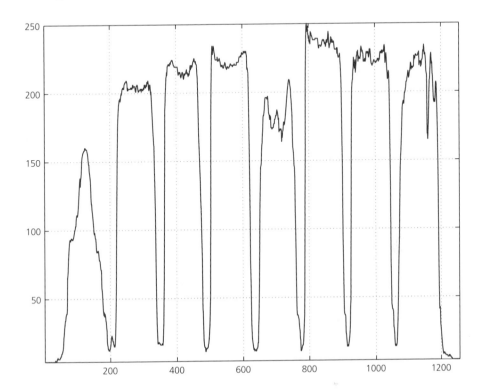

We can now see that the image consists of eight bands whose intensity varies from between 200 and 250 with the exception of the first one that has a lower intensity of around 100–150, and the fifth that is between 150 and 200. In all cases, the intensity of the background is around 20, so we can safely say that everything above 30 is a band and everything below is background.

We would now want to know the edges of these bands to know the positions of each one. To do this we need two intermediate steps. First we will set a threshold to create a 1D matrix with a 1 value for the bands and a 0 value for the background:

```
WesternBlot_BandBackground_C   = (WesternBlot_maxColumn>30);
WesternBlot_BandBackground_CD  = ...
     diff(WesternBlot_BandBackground_C);
figure
plot(WesternBlot_BandBackground_CD)
grid on    % add a grid to the display
axis tight % remove space of the plot to make the axis tight
```

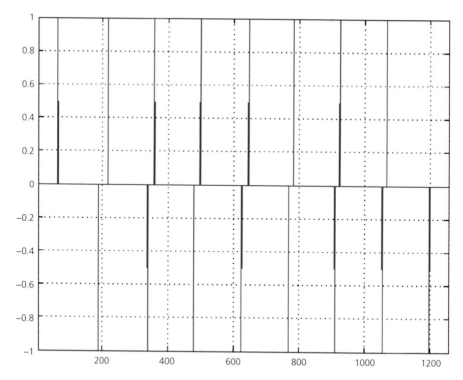

The previous figure shows a positive value (sometimes called spikes) at the columns where bands start, and a negative value at the columns where a band finishes. To get the positions of those spikes we use the command 'find' in the following way:

```
BandInitialColumn       = find(WesternBlot_BandBackground_CD>0);
BandFinalColumn         = find(WesternBlot_BandBackground_CD<0);

disp([BandInitialColumn;BandFinalColumn])
```

Columns 1 through 6

61	216	359	498	646	783
187	340	479	625	768	909

Columns 7 through 8

923	1066
1053	1197

We can see that the first band spans from column 61 to 187, the second from 216 to 340 and so on.

We are now interested in the rows at which the bands appear, but before we proceed, there is one other interesting fact that can be observed: the bands are not completely horizontal; there seems to be a slope. Since blots can be obtained in many ways that involve manual procedures, it is common to have results that have been rotated like this one. To have a better processing we can easily correct for this angle on the image by rotating the image to have a horizontal result. Rotations of a few degrees are sufficient in most of the cases, and this is obtained with the command 'imrotate' in the following way:

```
% rotate an image by 2 degrees clockwise
WesternBlot_2R  = imrotate(WesternBlot_2,-2,'crop');
figure
imagesc(WesternBlot_2R)
```

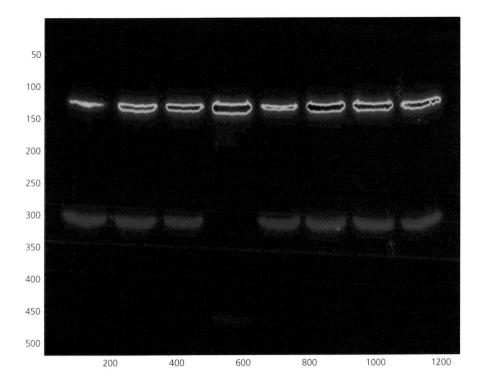

The command 'imrotate' will require as arguments, the image to be rotated and the rotation in degrees; positive numbers will rotate counter clockwise and negative numbers will rotate clockwise. The third and optional parameter will specify if we want the output image to be of the same dimensions as the input image ('crop') or if we want the output to be larger to accommodate the whole original image to be displayed and padded around it with zeros.

We can see that the bands look horizontal, but we can verify that 2 degrees clockwise is the optimum rotation by obtaining the maximum value of the rows as we did previously for the columns. We can scan every row of the original image to find its maximum value with 'max':

```
WesternBlot_maxRow          = max(WesternBlot_2,[],2);
```

Notice that the only difference is that we used a "2" instead of a "1" to specify the direction of the rows. Now to compare the results with those of the rotated images, we will do several rotations and compare the maxima with plots.

```
% rotate an image by 1 degrees counter clockwise
WesternBlot_2R_1pos      = imrotate(WesternBlot_2,1,'crop');
WesternBlot_maxRow_1pos = max(WesternBlot_2R_1pos,[],2);
% rotate an image by 1 degrees clockwise
WesternBlot_2R_1neg      = imrotate(WesternBlot_2,-1,'crop');
WesternBlot_maxRow_1neg = max(WesternBlot_2R_1neg,[],2);
% rotate an image by 2 degrees clockwise
WesternBlot_2R_2neg      = imrotate(WesternBlot_2,-2,'crop');
WesternBlot_maxRow_2neg = max(WesternBlot_2R_2neg,[],2);
% rotate an image by 3 degrees clockwise
WesternBlot_2R_3neg      = imrotate(WesternBlot_2,-3,'crop');
WesternBlot_maxRow_3neg = max(WesternBlot_2R_3neg,[],2);
```

Now that we have calculated the maximum value of each row with a series of rotations we can display each case. What we hope is that the rotation would have reduced the width of the band, which would indicate that this is the ideal rotation to compensate the orientation of the image. For this example we know that 2 degrees is the optimal rotation so we can highlight this line and the original one by increasing the width of the two lines.

```
figure
% display original image maximum in black
plot(WesternBlot_maxRow,'k','linewidth',2)
hold on
% display positive rotation in blue
plot(WesternBlot_maxRow_1pos,'b:')
% display negative rotation in red
plot(WesternBlot_maxRow_1neg,'r:')
% display negative rotation in red
plot(WesternBlot_maxRow_2neg,'r-','linewidth',2)
% display negative rotation in red
plot(WesternBlot_maxRow_3neg,'r--')
grid on
axis tight
legend('Original','Rotation +1',...
        'Rotation -1','Rotation -2','Rotation -3')
```

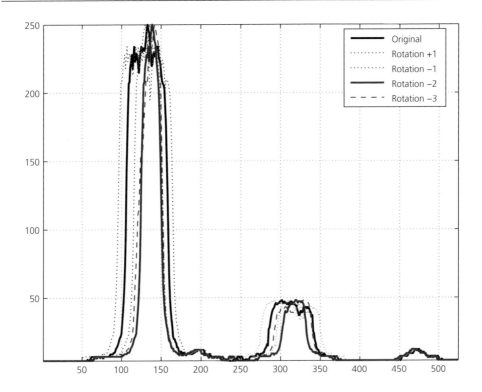

The plot reveals that the black line that corresponds to the original image is wider than the red lines, which then indicates that the rotation of the image concentrates the intensity into a smaller range of rows. The narrowest peak corresponds to the solid line of 2 degrees, with rotation 3 the band starts to get wider as it happens when a positive rotation is performed (blue dotted line).

To determine the extent of the bands we follow the same procedure as before, that is, threshold with a level to discriminate background from bands, calculate the difference and find the positions of the positive and negative values.

```
WesternBlot_BandBackground_R  = (WesternBlot_maxRow_2neg>30);
WesternBlot_BandBackground_RD = ...
   diff(WesternBlot_BandBackground_R);

BandInitialRow   = find(WesternBlot_BandBackground_RD>0);
BandFinalRow     = find(WesternBlot_BandBackground_RD<0);

disp([BandInitialRow BandFinalRow])
```

```
   119    156
   303    332
```

We can now see that the band from the upper part of the image spans between rows 119 and 156 and the lower one between 303 and 332. A very important observation arises here. From the plot, it can be seen that there is a smaller band around 450–500, which is generated by a very faint band of the blots. This band has a maximum intensity of 12, which is close to some of the values of the background close to the intense bands. Whenever a threshold value is selected, there is always a compromise. The lower it is, the more real signal it will select together with more noise. The higher it is, it will have the opposite effect; noise will be effectively removed at the expense of losing faint values of real data. There are techniques to find optimal thresholds, but they are more complicated, so for the time being, we will knowingly discard that faint band at the bottom of the image.

We have now located the positions of 16 different bands distributed in 2 rows and 8 columns. If we would like to extract some measurements of each of the bands, we would need to define their regions. We can first display the regions over the image to visualise the positions we are interested in.

To consider each region separately we need to evaluate them separately and to do that we will create two "loops", one that considers every position in the rows, and another that considers every position in the columns. By nesting one loop inside another we will evaluate every possible combination of rows and columns. Nested loops are common practice in programming.

We will now display the rotated Western Blot and use the two nested loops to draw boxes around each band region.

```
figure
imagesc(WesternBlot_2R_2neg)
for counterRows=1:numel(BandInitialRow)
    % Cycle over the rows has started.
    % Determine the positions of the rows for each
    % iteration of the loop
    RowPosition =[BandInitialRow(counterRows)  ...
                  BandFinalRow(counterRows)];
    for counterCols = 1:numel(BandInitialColumn)
        % Cycle over the columns has started
        % Determine the positions of the columns for each
        % iteration of the loop
        ColPosition = [BandInitialColumn(counterCols)  ...
                       BandFinalColumn(counterCols)];
        %Draw four lines to delineate a box at every band
        line(ColPosition,[RowPosition(1) RowPosition(1)], ...
            'color','w')
        line(ColPosition,[RowPosition(2) RowPosition(2)], ...
            'color','w')
        line([ColPosition(1) ColPosition(1)],RowPosition, ...
            'color','w')
        line([ColPosition(2) ColPosition(2)],RowPosition, ...
            'color','w')
    end
end
```

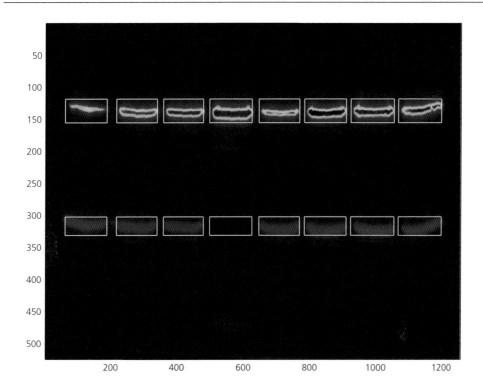

Once we have identified the regions of interest with the boxes, we can then extract some useful information and display it there. We have seen how to obtain means and maximum values, but we can have more than that. For instance we can obtain histograms of each band. We will again loop over every region and use a very similar code to the previous example. We will need some auxiliary variables to count each band.

```
currentBand =1;
for counterRows=1:numel(BandInitialRow)
    % Cycle over the rows has started.
    % Determine the positions of the rows for each
    % iteration of the loop
    % This time we will create a 1D matrix with all
    % the row positions
    RowPosition=(BandInitialRow(counterRows):...
            BandFinalRow(counterRows));
    for counterCols = 1:numel(BandInitialColumn)
        % Cycle over the columns has started
        % Determine the positions of the columns for each
        % iteration of the loop
        ColPosition =   (BandInitialColumn(counterCols):...
                BandFinalColumn(counterCols));
        % Store the pixels of the
        % current band in a temporal variable
```

```matlab
pixelsInsideBox   = double(WesternBlot_2R_2neg(...
                            RowPosition,ColPosition));
% Determine average, standard deviation
% and maximum values
% of the pixels inside the boxes:
meanValue           = mean(pixelsInsideBox(:));
stdValue            = std(pixelsInsideBox(:));
maxValue            = max(pixelsInsideBox(:));
```

```matlab
% display values on the image
averageText = strcat('Av:',num2str(round(meanValue)));
stdText     = strcat('Std:',num2str(round(stdValue)));
maxText     = strcat('Max:',num2str(round(maxValue)));
% display the information on the image
% with the segmented bands, offset the position
% so that the text is above the bands
text(ColPosition(1)-5,...
    RowPosition(1)-90,strcat('Band: ',...
    num2str(currentBand)),...
    'fontsize',9,'color','w');
text(ColPosition(1)-5,RowPosition(1)-70,...
    averageText,...'fontsize',9,'color','w');
text(ColPosition(1)-5,RowPosition(1)-50,stdText,...
    'fontsize',9,'color','w');
text(ColPosition(1)-5,RowPosition(1)-30,maxText,...
    'fontsize',9,'color','w');
```

```matlab
% find the histograms for each band
[yy(:,currentBand),xx(:,currentBand)] = ...
    hist(pixelsInsideBox(:),(0:15:255));
% increment the value of the current Band
currentBand = currentBand+1;
    end
end
```

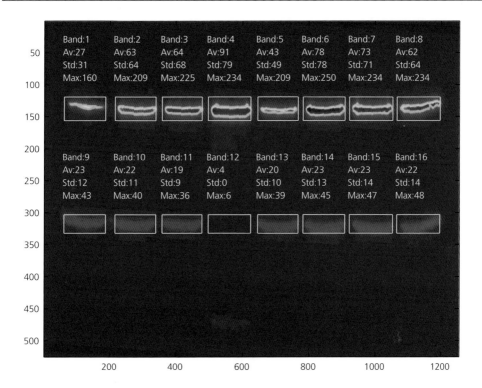

We can now identify each band and its characteristics. As we had mentioned initially, the fourth band from the left (bands 4 and 12) have very different characteristics, and we could assert that band 12 does not show any increase of intensity related to the presence of a protein. The average intensity of 4 and the maximum value of 6 may be that of the noise that forms the image. We can verify this by exploring the a section of background between rows 350 and 450:

```
mean(mean(WesternBlot_2R_2neg(350:450,:)))

max(max(WesternBlot_2R_2neg(350:450,:)))


    ans =

       4.0633


    ans =

       8
```

On the other hand, band 4 shows a very strong response with an average of 91 and a maximum of 234 making it the strongest band of the image. Once we have obtained these measurements, it is easy to explore them further as needed.

Matlab has many 3D visualisation plots with which we can display all the histograms together in a single figure. Let's try one called 'ribbon', which displays one line for each histogram previously obtained.

```
figure
ribbon(xx,yy)          % Ribbon displays lines for every
                       % row of a pair of matrices
axis tight             % Removes extra space from the axes
rotate3d on            % Enables the figure to be rotated with
                       %  the mouse
view(125,50)           % Sets an angle of view for the figure
xlabel('Bands of the W. B.','fontsize',14)
ylabel('Intensity level of the image','fontsize',14)
zlabel('Occurrence of the intensities','fontsize',14)
```

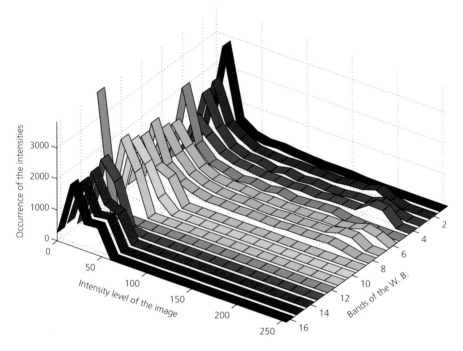

We can now see how some of the bands have many pixels in the bright region, i.e. the right side of the figure. Notice how band 4 has many pixels in the region of 200–250. On the other hand, band 12 has most of its pixels in the small numbers.

We can visualise more clearly the differences if we change the scale of the z-axis to be logarithmic. But before we can do that, there are many elements of the matrix 'yy' that contain zeros, and zeros cannot be displayed in a logarithmic plot (they correspond to minus infinity). One trick to work around this problem is to add one to the whole matrix, as the logarithm of one is zero. Moreover, if we add one to values of the order of thousands, it will not be a significant change.

```
figure
ribbon(xx,1+yy)
axis tight
rotate3d on
view(125,50)
set(gca,'zscale','log')
xlabel('Bands of the W. B.','fontsize',14)
ylabel('Intensity level of the image','fontsize',14)
zlabel('Occurrence of the intensities','fontsize',14)
```

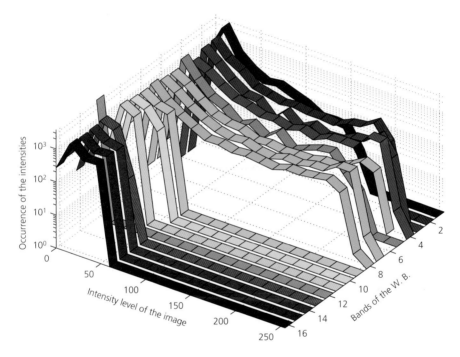

This final figure shows clearly how bands 1 to 8 have a considerable number of pixels with intensities above 60, bands 9–11 and 13–16 have lower intensities but still a considerable number above 30. Finally band 12 shows the majority of its pixels at very low levels, confirming no presence of protein in that band.

CHAPTER 5

Scratch Wound Assays

5.1 Analysis of Scratch Wound Assays

The analysis of in vitro cellular migration and proliferation through the well known "scratch wound assays" is a very popular tried-and-tested technique. Some elements of its popularity are its low cost and easy observation in which, upon the creation of a gap or wound, with a needle, micropipette or an electrical current, on a monolayer of cells, the cells will migrate or proliferate, or not, to close the wound and return to the state when the monolayer covered the whole area.

The importance of the analysis of cellular migration and proliferation in this type of "in vitro" setting is related to the understanding that the behaviour of these cells "in vitro" is mimicking the "in vivo" behaviour of the cells.

In this chapter, we will cover several techniques to analyse and quantify scratch wound assays to improve simple visual observations and save time from manual measurements.

We can start by downloading a typical image of a scratch wound assay. The image shows cells on the sides of the image and a uniform area in the middle. This is the area of the "wound".

```
migrationImage = imread(...
'http://caiman.group.shef.ac.uk/caiman/images/migration1.jpg');


figure
imagesc(migrationImage)
title('Example of a Scratch Wound Assay','fontsize',14);
```

Biomedical Image Analysis Recipes in MATLAB®: For Life Scientists and Engineers, First Edition.
Constantino Carlos Reyes-Aldasoro.
© 2015 John Wiley & Sons, Ltd. Published 2015 by John Wiley & Sons, Ltd.

Example of a scratch wound assay

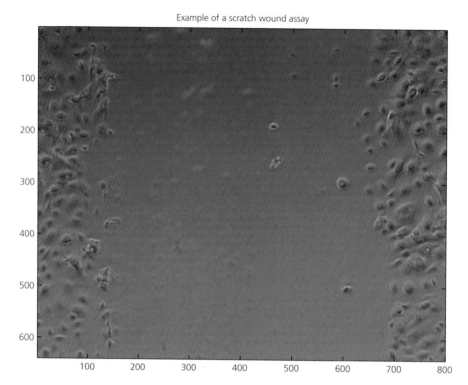

Image courtesy of Dr. Chryso Kanthou, The University of Sheffield, UK.

One usual way of analysing these type of images is to measure the size of the wound immediately after the monolayer has been "wounded", and then repeating the measurements after some time. The size of the wound would then be proportional to the phenomenon of interest, either migration or proliferation of the cells.

Whilst there are better ways to do this, it is possible to perform manual measurements on the image using Matlab's "image tool", which is invoked by typing `imtool(migrationImage)`. The command `imtool` has been described in more detail in the Western Blots chapter. The following image shows three measurements obtained with `imtool` in the example image. The measurements will show the different options that we can obtain manually. To avoid variability, some people decide to average a number of measurements, say five, which then increases the operator's time and the probability of errors from the users.

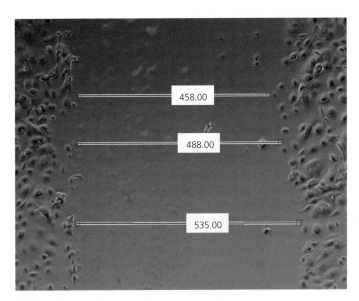

Before presenting any analysis techniques, it would be interesting to have a deeper understanding of the nature of the images. For this purpose, we will select a small region of interest and display two intensity profiles.

An area of interest can be selected by displaying the whole image and then zooming in by setting the visible region through the command 'axis', which is used to control the scaling and appearance of the axis on the figure. A suitable region of interest is around the two cells that appear next to each other and far from other cells in the upper right-hand section of the wound.

```
figure
imagesc(migrationImage)
axis([550 620 1 195 ])
```

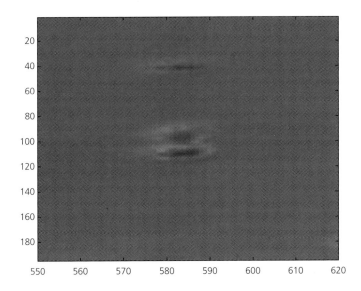

We can now select two intensity profiles in the region of interest. To keep the current image we use the command 'hold on' to be able to have several plots on the same figure, as the default is to erase whatever is present on a figure with a new plot.

```
hold on
plot([579 579],        [5 190],'k-', 'linewidth',2);
plot(586*ones(1,186),5:190 ,'r--','linewidth',2);
```

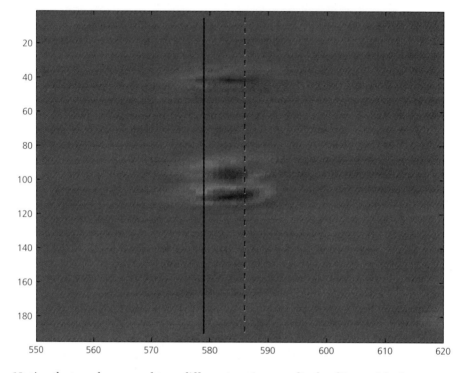

Notice that we have used two different options to display lines with the command 'plot'. In the first we defined the line with only two points; the two extremes of the line. The second command draws the line with many points, in this case 186 points, which correspond to one point for every unit between the extremes. For straight lines, both options are equivalent, but for lines other than straight lines it is necessary to define all points, or at least a number of them.

Now we can display the intensities of the profiles on a separate image. This is done again with the command 'plot' and selecting the rows and columns, which correspond to the intensity profiles indicated in the previous figure. We have used different line styles (solid, dashed) and colours (black, red) to distinguish each profile.

```
figure
plot(5:190,migrationImage(5:190,579,1),'k-',...
    5:190,migrationImage(5:190,586,1),'r--','linewidth',2)
grid on
axis tight
title('Two Intensity Profile from a R.O.I','fontsize',14);
```

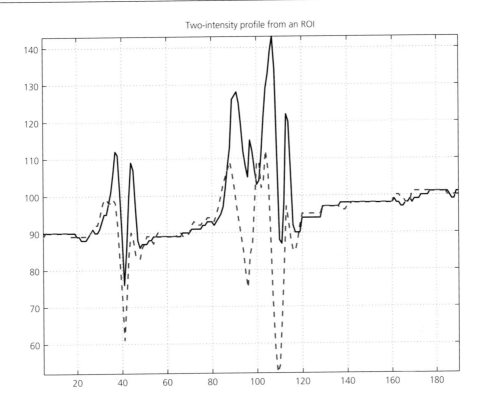

Two-intensity profile from an ROI

There are several interesting observations that arise from the profiles. First, the background of the image occurs at an intermediate level of grey, and the intensity corresponding to the cells is manifest as higher (bright) or lower (dark) levels. This is a common characteristic of numerous imaging modalities such as interference contrast images. A second observation is that the intensity of the background appears to increase gradually towards the right side of the plot. This is due to a problem of "shading" or the uneven intensity of the background, which is present in images of many acquisition techniques, from microscopy to Magnetic Resonance Imaging. Shading can be a result of several factors, such as an uneven illumination, different thickness of samples or uneven magnetic fields. Shading correction is treated in detail in the chapter of Bright Field Microscopy and will not be further discussed here.

One final observation is that the cells are defined by a sudden change of intensity, increasing and decreasing over the background. It is these variations that we will exploit over the following recipes to process and analyse the measurements of interest that we can extract of the scratch wound images.

5.2 Recipe 1: Low Pass Filtering Scratch Wound Assays in the Spatial Domain

In this recipe we will present one of a series of well-known techniques in which we will modify the intensity of the pixels of the image, based on the intensities of the pixel itself and several of its neighbours. Since the neighbours are limited in distance or 'space', these techniques are considered as spatial techniques or techniques in the spatial domain. If the signals correspond to measurements in time, instead of position or space, they are known as time domain techniques.

One simple modification that we can perform is to average a number of neighbours, and replace the value of each pixel with the average of its neighbours. There are many ways to do this in Matlab; we could add copies of a signal each with different points or address different regions, i.e. '(A(1:end-1) + A(2:end))/2'. In this simple example we are averaging every pair of neighbours together. The averaging gives the same weight to each element, therefore we could abstract this operation by indicating the weights that we use in a matrix, i.e. [0.5 0.5] or equivalently [1 1]/2.

While this is a quick and simple way to average neighbouring pixels, there are more efficient ways to combine neighbours by using one matrix to define the way we want to combine the neighbouring pixels; we will call this matrix a "filter". For instance, we can define a 3 × 3 filter in which we give equal weight to each element in the following way:

```
lowPassFilter_1           = [1 1 1;1 1 1;1 1 1]/9;
```

or equivalently:

```
lowPassFilter_1           = ones(3)/9;
```

We have called this filter "low pass" for reasons that will become apparent later. We have divided the filter over 9 so that the sum of all the elements of the filter is equal to 1.

We can then use a mathematical function called "convolution" to obtain the average of every 3 × 3 neighbourhood of pixels along the whole image. Matlab calculates convolutions with the command 'conv' for one dimension, 'conv2' for 2D and 'convn' for 3D and greater dimensions. The commands take the image (or signal) to be "convolved" as first argument and the filter as second argument. We can use an optional third parameter to indicate that we want to keep the same dimension of the result as the original image.

```
migrationImage_LPF        = conv2(...
double(migrationImage(:,:,1)), lowPassFilter_1,'same');
```

Notice that we have selected just one channel of the image; as for a grey-scale image the three channels provide the same information. Notice as well that we have used the command 'double' as the image is formed by unsigned integers and these are not used in mathematical operations like the convolution.

To show the effects of the filtering, we can display the profile of the original image together with the profile of the filtered image:

```
figure
plot(5:190,migrationImage(5:190,586,1),'k-',...
    5:190,migrationImage_LPF(5:190,586,1),'r--','linewidth',2)
title(['Original Profile (solid black);',...
       'Low Pass Filtered Profile [3×3] (dashed red)'],...
       'fontsize',14);
grid on
axis tight
```

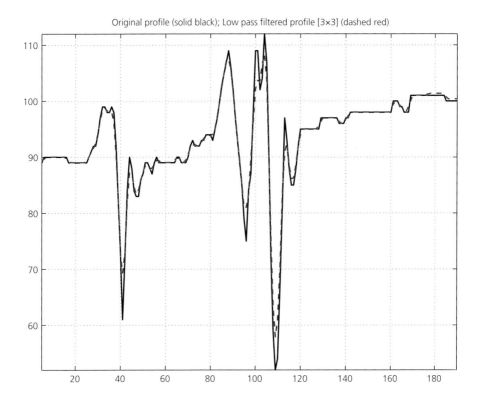

Original profile (solid black); Low pass filtered profile [3×3] (dashed red)

Notice that we have written the title in two instruction lines. When we want to break a line and continue in the next instruction line, we use three consecutive dots (...) to indicate that the line will continue. In addition, when we are breaking a string that is written in between single quotes, we need to contain each section in between single quotes and the square brackets '[ ]' at the beginning and the end of the string.

We can see that the two profiles closely follow each other except in the abrupt changes, for instance, at the narrow valley at the centre (coordinates 109,60). We can create a larger filter to have a stronger filtering effect. We could have [5 × 5] matrix with ones, but another option is to give a higher weight to pixels that are closer to the pixel to be filtered and a lower weight to pixels that are further away, for instance, filter like this [0.1 0.2 0.4 0.2 0.1]. The shape of this filter, symmetrical, with tails that decrease to the edges is a well known bell curve that can be created with a Gaussian function. Matlab provides a function to create Gaussians as well as many other common shapes of this sort through the command 'fspecial'. We can use this command to create our filters by indicating the size of the filter (say 15 × 15 for a bigger filter) and the width of the Gaussian function. For example this filter [0.05 0.2 0.5 0.2 0.05] is narrower, or has a lower standard deviation, than the previous example.

We can now create a second filter using fspecial:

```
lowPassFilter_2              = fspecial('Gaussian', [15 15],3);
```

We can now filter our original image and compare the same profiles as before:

```
migrationImage_LPF2 = imfilter(migrationImage(:,:,1),...
                      lowPassFilter_2,'replicate');
figure
plot(5:190,migrationImage(5:190,586,1),'k-',...
    5:190,migrationImage_LPF(5:190,586,1),'r-',...
    5:190,migrationImage_LPF2(5:190,586,1),'b:','linewidth',2)
grid on
axis tight
title(['Original Profile (solid black);',...
    ' [3×3] L.P.F. (dashed red);',...
    ' [15×15] L.P.F (dotted blue)'],...
    ' fontsize',12);
```

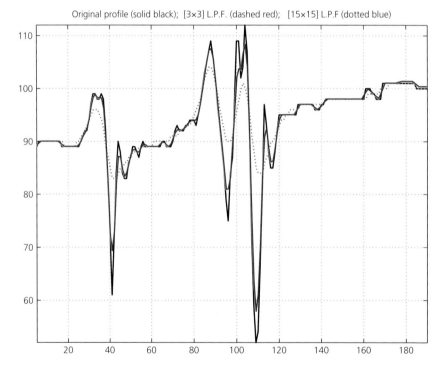

The effects of the filter are now clear; the sharp peaks of the profile have been eroded leaving a smoother version of the data. We can now observe the effect on the whole image. To appreciate the differences with the original image, we can display half of the filtered image next to half of the original image.

```
figure
imagesc([migrationImage_LPF2(:,1:400,1) migrationImage...
       (:,401:800,1) ])
colormap(gray)
title('Left:Low Pass Filtered,      Right: Original',...
    'fontsize',14);
```

Left: Low pass filtered, Right: Original

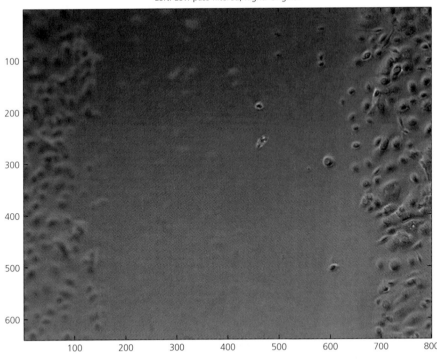

The smoothing effect is also visible as a blurred effect. In summary, a filter that averages values of neighbouring pixels will be removing fine detail of abrupt variations of intensity; this is useful to remove noise from images, but it may also remove detail in some cases.

Sometimes, it is convenient to close the figures that we have opened so far. This can be done with the command 'close'. If we type 'close' it will close the current figure. We can also close a specific figure e.g. 'close 1', or to close all figures we type:

```
close all
```

Conversely, sometimes we might want to remove some variables from the Workspace. We can do that with the command 'clear' followed by the name of the variable. When we want to clear all variables we type:

```
clear all
```

5.3 Recipe 2: High Pass Filtering Scratch Wound Assays in the Spatial Domain

The complementary operation of the low pass filter (LPF) is the high pass filter (HPF). Again we will apply a mathematical operation to the intensities of a reduced neighbourhood of pixels, but instead of aiming to average the values and smooth the peaks, we will try to find those abrupt changes of intensity. To do this, our filter will be designed to return a zero for neighbourhoods of uniform intensity and highlight areas where there is an abrupt transition of intensities. There are many ways in which this can be obtained, for instance by using a filter like this $[-0.5\ 0.5]$. For two dimensions, a good choice of filter is one that compares a central pixel with its neighbours to the sides, or more specifically that subtracts the value of the sides, left/right, up/down, from the value of the central pixel, for instance:

```
highPassFilter              = [0 -1 0;-1 4 -1; 0 -1 0];
```

We can use this filter with the same image of a scratch wound assay as in the previous recipe:

```
migrationImage = imread(...
'http://caiman.group.shef.ac.uk/caiman/images/migration1.jpg');
```

In the previous recipe we used the convolution operation to filter the image, and we noted that since the image was formed of unsigned integers it was necessary to convert these to doubles. Now we will introduce a different command that is specific for filtering images, the command 'imfilter', which stands for image filter. Since this command is designed for images, it works for integers as well as doubles and other data types. It also has more options that may be useful. For example, one important issue may arise on the edges of an image; when we want to apply the filter to the edge of the image, we cannot filter the pixel of coordinates $(1,1)$ as there are no pixels to the left and above this pixel. If the filter is relatively big, say 15×15, again we would not be able to process the pixels of the first 14 rows and columns. The command 'imfilter' provides different ways to process the boundaries of the image: by default it will not compensate and the pixels of the boundary are not affected by the filter. However, it can also pad the edge of the image with enough pixels so that the pixels of the boundary can be processed with the new padded neighbours. These surrounding pixels may be set to zero, or more conveniently, to the same values that are present on the edges. We obtain this padding by specifying the third parameter of imfilter as 'replicate'; the other two input arguments are the image to be filtered and the filter itself. The corresponding instruction to apply the high pass filter to the image follows:

```
migrationImage_HPF   = imfilter(migrationImage(:,:,1),...
                       highPassFilter,'replicate');
```

We can now display the profiles to analyse the effects of the filter:

```
figure
plot(5:190,migrationImage(5:190,586,1),'k-',...
    5:190,80+migrationImage_HPF(5:190,586,1),'r--',...
    'linewidth',2)
```

```
title(['Original Profile of R.O.I (solid black);',...
    'High Pass Filtered Profile [3×3] (dashed red)'],...
    'fontsize',14);
grid on
axis tight
```

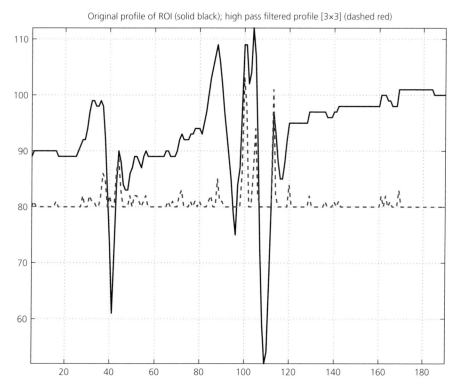

Original profile of ROI (solid black); high pass filtered profile [3×3] (dashed red)

The profiles show very clearly the effect of a high pass filter: the filter responds to sudden changes of intensity, either increasing or decreasing. Notice how the filter detects even the very subtle changes on the left of the figure between the 0 and 20. On the other hand, where the intensity is constant there is no response of the filter. Notice that we have shifted upwards the response of the filter by adding a constant of 80; this is solely to be able to compare the two signals. An alternative way to display two signals that have different intensity ranges is to replace the command 'plot' with 'plotyy', which is used to display two signals with different y-axes, thus the 'yy'. The following command repeats the previous figure with plotyy:

```
figure
plotyy(5:190,migrationImage(5:190,586,1),...
    5:190,migrationImage_HPF(5:190,586,1))
title(['Original Profile of R.O.I (blue);',...
    'High Pass Filtered Profile [3×3] (green)'],...
    'fontsize',14);
grid on
```

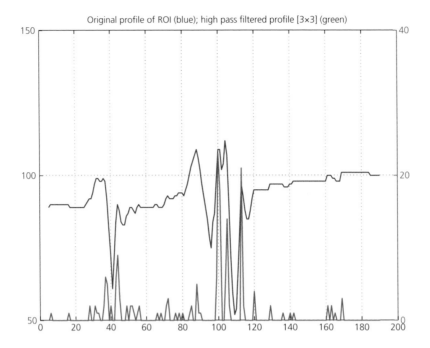

Original profile of ROI (blue); high pass filtered profile [3×3] (green)

We can now display the filtered image next to the original image. Again we will add a constant value of 80 so that the range of grey levels of the filtered image is the same as the original image. Furthermore, we can scale the filtered image to enhance the visual observation of the effects of the filter. For comparison purposes we will display the original image next to the filtered image (a) without scaling nor shifting, (b) shifted by a constant of 80 and (c) shifted by 80 and scaled by a factor of 3:

```
figure
imagesc([migrationImage_HPF(:,1:400)...
    migrationImage(:,401:800,1) ])
colormap(gray)
title('Left: (a) High Pass Filtered,        Right: Original',...
    'fontsize',14);
figure
imagesc([80+migrationImage_HPF(:,1:400)...
    migrationImage(:,401:800,1) ])
colormap(gray)
title(['Left: (b) High Pass Filtered shifted,        Right:',...
    'Original'],'fontsize',14);
figure
imagesc([80+3*migrationImage_HPF(:,1:400)...
    migrationImage(:,401:800,1) ])
colormap(gray)
title(['Left: (c) High Pass Filtered shifted and scaled,',...
    'Right: Original'],'fontsize',14);
```

Left: (a) High pass filtered, Right: Original

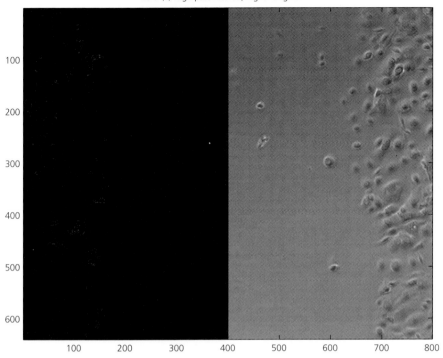

Left: (b) High pass filtered shifted, Right: Original

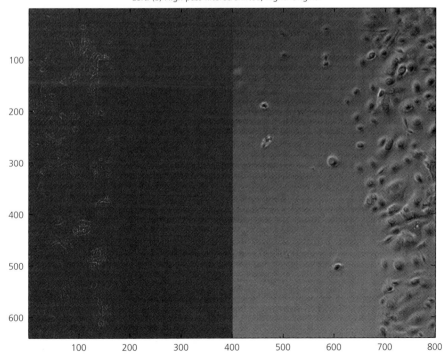

Left: (c) High pass filtered shifted and scaled, Right: Original

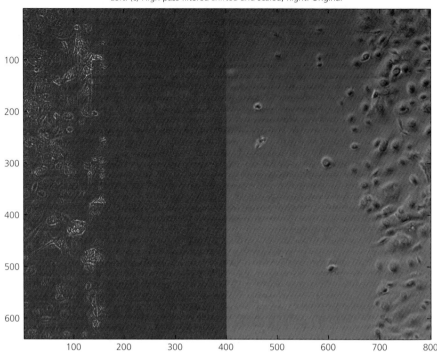

One characteristic of high pass filters is that they react strongly to small changes of intensity, that is, noise can be amplified. A second interesting observation is on the uniformity of the intensity. We had previously observed that this image was increasing in intensity along the profiles. Since the high pass filters react to sudden changes of intensity and not to constant, or very smooth changes, the background of the filter response remains constant along the profile. This is also evident in the image comparison, where, the top the grey level of the image is similar to the shifted filtered image, and at the bottom, the image is far brighter.

The previous filter was a small [3 × 3] filter. We saw in the previous recipe that larger filters may have different effects; therefore it is interesting to know what the effect of a larger filter would be. We will analyse this by taking a small shortcut, that is, exploiting the properties of the filters.

While the high pass filters react to sudden variations of the image, the low pass filters react to the slow variations of the image; therefore they have complementary effects. We can apply a high pass filter to an image with a filter based on the previous low pass filter result and vice versa. We can illustrate this complementarity by repeating the low pass filter of the previous recipe:

```
lowPassFilter_2          = fspecial('Gaussian', [15 15],3);
migrationImage_LPF2      = imfilter(migrationImage(:,:,1),...
                           lowPassFilter_2,'replicate');
```

With this result we can now obtain the high pass filtered version of the image simply by subtracting the filtered image from the original image:

```
migrationImage_HPF2      = (double(migrationImage(:,:,1))-...
                           double(migrationImage_LPF2));
```

Notice that we converted both images to doubles since unsigned integers do not allow negative numbers. We can now display the profile and image comparison in the same way as before.

```
figure
plot(5:190,migrationImage(5:190,586,1),'k-',...
     5:190,80+migrationImage_HPF2(5:190,586,1),'r--',...
     'linewidth',2)
grid on
axis tight
title(['Original Profile of R.O.I (solid black);',...
      ' High Pass Filtered Profile [15x15] (dashed red)'],...
      'fontsize',14);

figure
imagesc([80+migrationImage_HPF2(:,1:400)...
     migrationImage(:,401:800,1) ])
colormap(gray)
title(['Left: Original - Low Pass Filtered,       Right:',...
     'Original'],'fontsize',14);
```

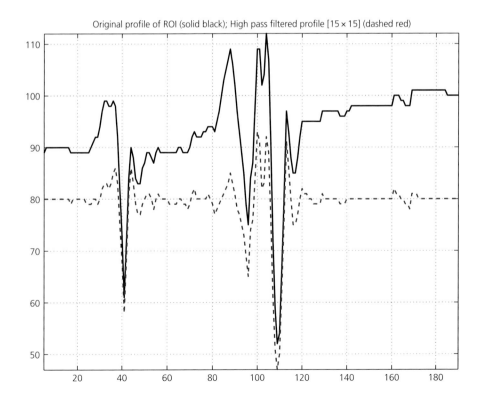

Original profile of ROI (solid black); High pass filtered profile [15 × 15] (dashed red)

Left: Original – Low pass filtered, Right: Original

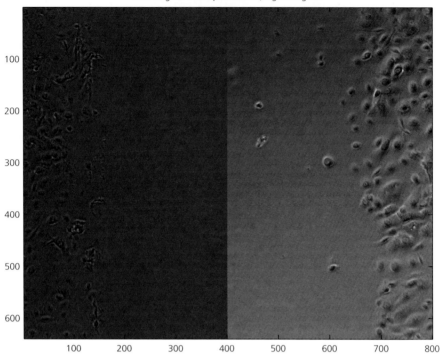

The results show the effect of the high pass filtering with a larger filter, the cells on the right are smoother than the first high pass filter and we still have the effects of a uniform background as before.

Before we finish this recipe, we need to highlight the fact that filters come in many shapes, and they provide many different effects. For instance, a popular shape is the "Laplacian of a Gaussian" (LOG), which looks a bit like an inverted "Mexican Hat". This filter has interesting properties that we will investigate by filtering our example image.

To create the filter we can use 'fspecial' again, and instead of specifying a Gaussian function we specify 'log', which stands for Laplacian of a Gaussian and use that to filter the image in the same way as before:

```
highPassFilter_3              = fspecial('log', [15 15], 2);
migrationImage_HPF3           = imfilter(migrationImage(:,:,1),...
                                highPassFilter_3,'replicate');
```

We will display the filtered profile shifted and scaled to appreciate the effects of the filter.

```
figure
plot(5:190,migrationImage(5:190,586,1),'k-',...
```

```
    5:190,80+10*migrationImage_HPF3(5:190,586,1),'r--',...
    'linewidth',2)
grid on
axis tight
title(['Original Profile of R.O.I (solid black);',...
    ' High Pass Filtered Profile L.O.G. [15x15]',...
    '(dashed red)'],...
    'fontsize',14);
```

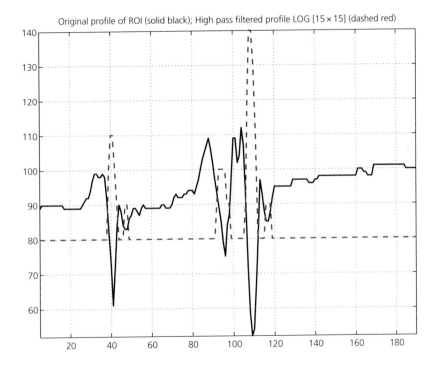

The Laplacian of a Gaussian filter has interesting properties; it does not respond to regions of uniform intensity either to small increasing or decreasing variations. It reacts only to regions where the intensity drops and recovers abruptly, like a deep narrow trough.

The corresponding image shows the effects of this filter in two dimensions.

```
figure
imagesc([80+20*migrationImage_HPF3(:,1:400,1)...
        migrationImage(:,401:800,1) ])
colormap(gray)
title(['Left: Filtered with Laplacian of Gaussian,',...
    'Right: Original'],'fontsize',14);
```

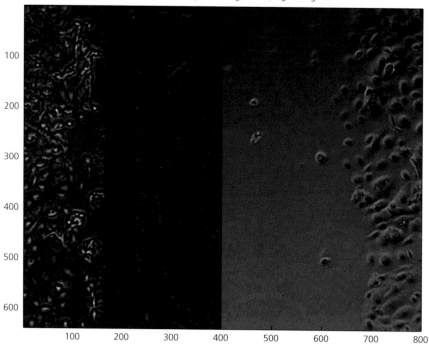

Left: Filtered with laplacian of gaussian; Right: Original

If we display the four filters analysed so far, we might have a better understanding of the effects they produce.

```
lowPassFilter_1                = [1 1 1;1 1 1;1 1 1]/9;
figure

subplot(221)
surf(lowPassFilter_1)
title('Low Pass Filter [3×3]');
axis tight

subplot(222)
surf(lowPassFilter_2)
title('Low Pass Filter [15×15]');
axis tight

subplot(223)
surf(highPassFilter)
title('High Pass Filter [3×3]');
axis tight

subplot(224)
surf(highPassFilter_3)
title('Laplacian of Gaussian Filter [15×15]');
axis tight
```

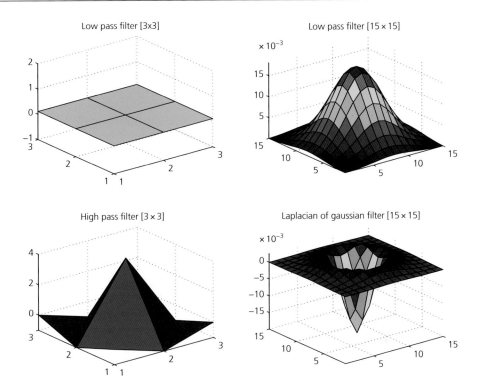

The low pass filters of the top row have only positive values and therefore will always create an average, or a weighted average, of the values of the pixels of a neighbourhood. The high pass filters on the other hand have positive and negative values, and therefore will produce high values where the shapes fit to regions of the image. That is, a speckle of noise looks like the [3 × 3] high pass filter, and the troughs that fit with the inverted hat of the [15 × 15] filter will also have a strong response.

Finally we can compare the four filtered images corresponding to the four filters of the previous figure, in this case, without the original image on the right hand side.

```
migrationImage_LPF  = conv2(double(migrationImage(:,:,1)),...
                     lowPassFilter_1,'same');
figure

subplot(221)
surf(lowPassFilter_1)
imagesc(migrationImage_LPF)
colorbar
title('Low Pass Filter [3x3]');
axis tight

subplot(222)
imagesc(migrationImage_LPF2)
title('Low Pass Filter [15x15]');
colorbar
axis tight
```

```
subplot(223)
imagesc(migrationImage_HPF)
title('High Pass Filter [3×3]');
colorbar
axis tight

subplot(224)
imagesc(migrationImage_HPF3)
title('Laplacian of Gaussian Filter [15×15]');
colorbar
axis tight

colormap(gray)
```

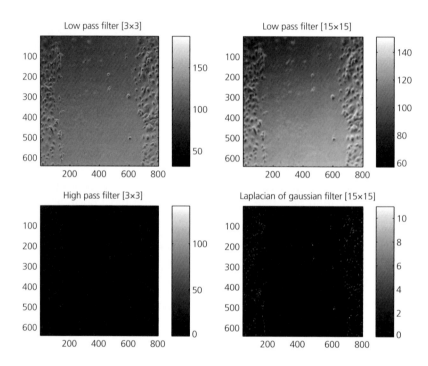

The 'colorbar' reveals interesting details of the image, for instance, the larger low pass filter has reduced the range of intensities from approximately [40–180] to [60–150]. Conversely, the response of the [3 × 3] high pass filter spans a very large range compared with the [15 × 15] Laplacian of a Gaussian.

5.4 Recipe 3: Combining Filters and Morphological Operations

In the previous recipes we have seen the effects of two basic type of filters; low pass filters smooth images by averaging elements in a small neighbourhood, and high pass filters have the opposite effect of highlighting abrupt changes in intensity. In this recipe, we will see how we can combine the effects of several filters to detect the cellular regions of a scratch wound assay. We will be able to detect small regions or use other operations to generalise and find a single region, which corresponds to the wound with two cellular regions to the sides.

We can begin with the same image as in the previous recipes:

```
migrationImage = imread(...
'http://caiman.group.shef.ac.uk/caiman/images/migration1.jpg');
```

We will first define two filters. First, a [3 × 3] high pass filter is used to find areas of the region where there is sharp variation of the intensities. By doing so, we will discard any problems of shading of the original image. We will later use a [15 × 15] Gaussian low pass filter. The effect of this filter will be to smooth the results of the first HPF. The filters are defined like this:

```
highPassFilter        = [0 -1 0;-1 4 -1; 0 -1 0];
lowPassFilter_2       = fspecial('Gaussian', [15 15],3);
```

We again use the command 'imfilter' to obtain the high pass and low pass filtered versions over one channel of the original image.

```
migrationImage_HPF        = imfilter(migrationImage(:,:,1),...
    highPassFilter,'replicate');
migrationImage_HPF_LPF    = imfilter(abs(migrationImage_HPF),...
    lowPassFilter_2,'replicate');
figure
imagesc((migrationImage_HPF_LPF))
colorbar
title('Filter Concatenation: HPF -> LPF','fontsize',14);
colormap(gray)
```

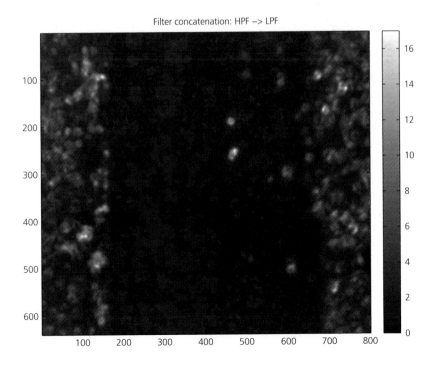

Filter concatenation: HPF –> LPF

The combination of the two filters provides an interesting result. The high pass filter first highlights all the regions where there are sharp variations of the intensity, then the low pass filter will average the neighbours and the result is an image with a uniform background in which the intensity is a reflection of the presence of cells in the original image.

It is very important to notice that the order of the filters matters. To illustrate this point, we can invert the order and do the low pass first and then the high pass filtering.

```
migrationImage_LPF        = imfilter(migrationImage(:,:,1),...
                            lowPassFilter_2,'replicate');
migrationImage_LPF_HPF    = imfilter(abs(migrationImage_HPF),...
                            highPassFilter,'replicate');
figure
imagesc((migrationImage_LPF_HPF))
title('Filter Concatenation: LPF -> HPF','fontsize',14);
colormap(gray)
```

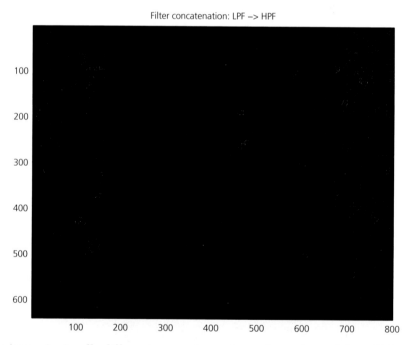

Filter concatenation: LPF –> HPF

The image is visually different, as we do not have the regions of the cells but rather points of high intensity variation.

The next step in the analysis of the scratch wound assay is to assign a threshold to the intensity of the filtered image (HPF-LPF) to determine the regions with and without cells. This is called "segmentation by thresholding". We could assign the threshold manually, but Matlab provides a function called 'graythresh' to determine the optimal threshold for grey-scale images. The command 'graythresh' uses a very established algorithm proposed by Otsu in 1979. The algorithm assumes that the data is distributed in two classes of pixels, which form a foreground and a background. The result of 'graythresh' is between 0 and 1 so we need to scale it to be between 0 and 255 to apply it to our image.

```
thresLevel     = 255*graythresh((migrationImage_HPF_LPF));
cellularRegion = migrationImage_HPF_LPF>thresLevel;
```

At this moment we can start obtaining quantitative measurements from the results obtained. Two interesting measurements are the area of the image that is covered by cells (or background) and the number of regions of connected pixels, e.g. the regions of isolated cells and regions where the segmentation has merged several cells into a single region. We are interested in the number of regions. The area is easily quantified from the binary image by using the command 'sum' as it will add the values of all pixels inside the image; in this case, the values will be either 0 for background or 1 for cells. Since this measurement would depend on the size of the image, it is better to find the relative area of the cells, that is, the number of pixels corresponding to the area divided by the dimensions of the image.

```
[rows,columns,levels]   = size(migrationImage);
areaCoveredbyCells      = sum(cellularRegion(:))/rows/columns;
```

To measure the number of connected regions of pixels, we need to first identify each of them; in other words, we need to assign a unique label to each of them. Matlab assigns labels and counts the number of labels with the command 'bwlabel' for 2D images and 'bwlabeln' for data with higher dimensions.

```
[cellularRegion_Labelled,numRegions] = ...
    bwlabel(cellularRegion);
figure
imagesc(cellularRegion_Labelled)
title('Labelled Cellular Regions','fontsize',14);
```

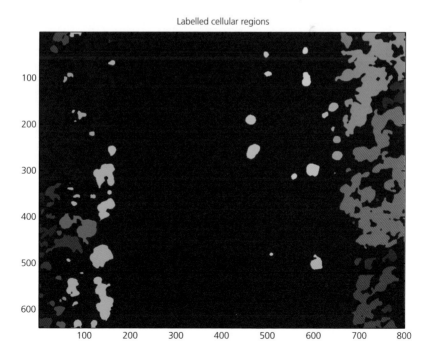

The previous image is displayed with a different colour for every labelled region. The labelling process scans the image from the top left, going down first and then to the right in the process of assigning labels. Therefore, the colours increase down and right as we see in the labelled image. In some cases, it is convenient to have these colours arranged in sequence. However, in some cases it would be better to have neighbouring regions with different colours so that they are easier to distinguish visually. One easy way to do this is to change the colours of the image by changing the colour map. We can use some predefined colour maps like 'hot' or 'hsv', but we can also create our own colour maps. One simple example is to assign colours randomly using the command 'rand' that generates random numbers. We need to create a matrix of 3 columns and a certain number of rows, for instance 64.

```
colormap(rand(64,3))
```

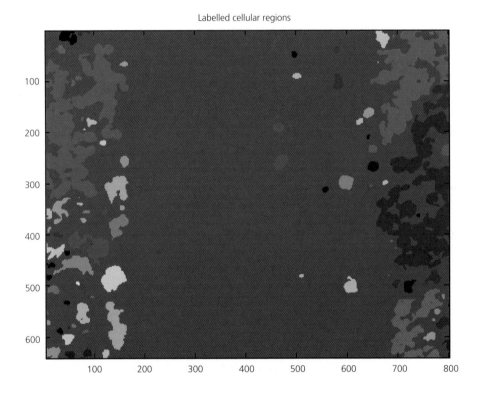

Labelled cellular regions

Since the colours are randomly assigned, every time we assign a colour map it will be different. Another way to change the colours is to use those of a predefined colour map. For instance, we can use the one that Matlab uses as default, which is called 'jet' and spans the colours of the rainbow. Jet is a matrix of 64 rows and 3 columns, so we will copy that matrix to a new matrix. Next we will create and sort a matrix of 64 values so that we can have an index with which we will reorder, randomly, the colours of the 'jet' rainbow. Finally we can set the first row, that which corresponds to the value 0, to be black so that the background is black and we have a better contrast between the regions and the background. We can create that effect with the following lines:

```
jetColours                    = jet;
[sortedValues,indexRandomValues]        = sort(rand(64,1));
randomJetColours              = (jetColours(indexRandomValues,:));
randomJetColours(1,:)     = [ 0 0 0];
colormap(randomJetColours)
```

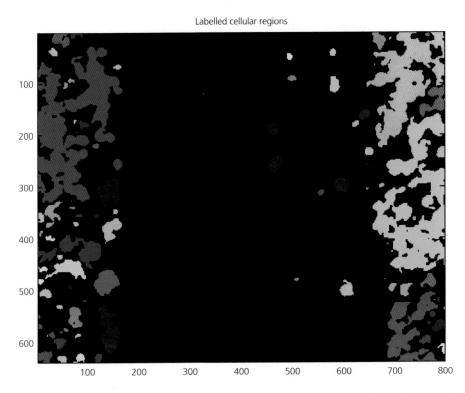

Labelled cellular regions

Since this change of colour map is something that may be done frequently, it is useful to save these lines as a shortcut. For an explanation of shortcuts refer to Chapter 1.

We can now display the cellular image in a different way by overlaying the thresholded binary image on the original image. This will allow us to visually examine the results of the segmentation. To overlay the results, we create a copy of the original image, and then will add the results to one of the channels, in this case the second channel so that the image will have a grey shade for the background and a green colour for the cells. Since we have already calculated the number of regions as well as the area covered by the cells we can include that in the title of the image.

```
migrationImage_res            = migrationImage;
migrationImage_res(:,:,2) = migrationImage_res(:,:,2)+...
                          50*uint8(cellularRegion);
migrationImage_res(migrationImage_res>255) = 255;
```

```
figure
imagesc(migrationImage_res)
title(strcat('Cellular Region = ',...
    num2str(100*areaCoveredbyCells),...
    ' [%];    Num Regions = ',...
    num2str(numRegions)),'fontsize',14)
```

Cellular region = 17.8568 [%]; Num regions = 94

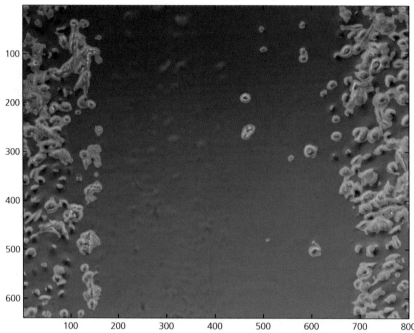

Notice that we had to convert the cellular region from a 'logical' data type to 'uint8' in order to overlay the segmentation on the image. The value of +50 implies that the pixels will have a larger than normal (i.e. grey) value and will thus stand out as green. The following line is used to restrict the range of values to 255 as the previous addition could lead to values higher than the maximum for a type of 'uint8'. We could also subtract a value and that would bias the shade to the opposite of green. In that case we need to verify that no value is lower than zero:

```
migrationImage_res            = migrationImage;
migrationImage_res(:,:,2)     = migrationImage_res(:,:,2)...
                                -50*uint8(cellularRegion);
migrationImage_res(migrationImage_res<0)  = 0;
figure
imagesc(migrationImage_res)
title(strcat('Cellular Region = ',...
    num2str(100*areaCoveredbyCells),...
    '[%];    Num Regions = ',num2str(numRegions)),...
    'fontsize',14)
```

Cellular region = 17.8568 [%]; Num regions = 94

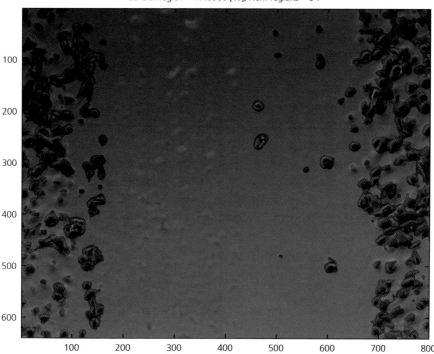

Adding or subtracting the results into other channels would provide different shades. A visual examination of the results shows that the segmentation is rather good, as the background is well segmented from the cells.

5.5 Recipe 4: Sensitivy to Thresholds and Hysteresis Thresholding

In the previous recipes we have seen how we can combine filters to segment cells in a scratch wound assay. The results that we obtained were good, however, there is one very important parameter that was used to obtain those results, e.g. a threshold level was used to distinguish the background from the cells. Whenever we use a single parameter that may influence the results, it is important to analyse the impact of that parameter, or in other words, the sensitivity to this parameter. One way of analysing this sensitivity is to vary this parameter, in this case the threshold, and observe how it impacts the results. In our case, we can set threshold levels from the minimum level of intensity to the maximum. In this way we will cover every possible outcome. Of course, the two extremes are trivial cases. As with the minimum level the entire image will be covered and with the maximum most of the image will be considered as background.

```
migrationImage = imread(...
'http://caiman.group.shef.ac.uk/caiman/images/migration1.jpg');
```

```
highPassFilter              = [0 -1 0;-1 4 -1; 0 -1 0];
lowPassFilter_2             = fspecial('Gaussian', [15 15],3);
migrationImage_HPF          = imfilter(migrationImage(:,:,1),...
                                    highPassFilter,'replicate');
migrationImage_HPF_LPF = imfilter(abs(migrationImage_HPF),...
                                    lowPassFilter_2,'replicate');

minIntensity                = min(double(migrationImage_HPF_LPF(:)));
maxIntensity                = max(double(migrationImage_HPF_LPF(:)));

disp([minIntensity maxIntensity])
    0    17
```

For our example the minimum and maximum levels are 0 and 17, therefore we will create a matrix that spans these values. We can create this matrix in several ways; one is by using the command 'linspace' that creates a linear space with a number of points from an initial to a final value.

```
variableThreshold                   = linspace(minIntensity,...
                                    maxIntensity,18);
```

We can now use a "loop" and apply the different thresholds. We will save all intermediate results, so the thresholded images will be stored in a 3D matrix, and the area and number of regions will be two separate 1D matrices.

```
[rows,columns,levels]               = size(migrationImage);

for counterThres= 1:18
    cellularRegion_thres(:,:,counterThres)    = ...
      migrationImage_HPF_LPF>=variableThreshold(counterThres);
    areaCoveredbyCells(counterThres)          = ...
      sum(sum(cellularRegion_thres(:,:,counterThres)))...
      /rows/columns;
    [celRegion_Lab,numRegions(counterThres)]= ...
        bwlabel(cellularRegion_thres(:,:,counterThres));
end
```

We can now display the areas and number of regions that are generated with the variable thresholds. We will display both cases in a single graph. Since the area is a number between 0 and 1 we can scale that to be a percentage by multiplying it by 100.

```
figure
plot(variableThreshold,100*areaCoveredbyCells,'b-o',...
    variableThreshold,numRegions,'r-x','linewidth',2);
grid on
legend('Area Covered by Cells','Number of Regions Segmented');
xlabel('Threshold level','fontsize',14)
ylabel('Area [%] / Number Regions','fontsize',14)
```

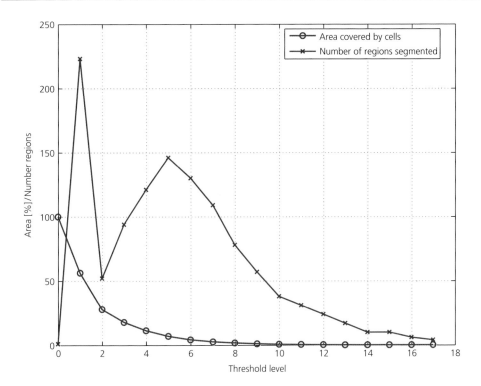

The results are quite interesting. The area decreases with the increase of the threshold from 100 to 0 as it was expected. However, the number of regions present a rather strange behaviour. With the threshold at zero, there is a single region. With the maximum threshold the regions drop to zero. However, the transitions in between show a peak at 1, which drops to 2 and then there is a gradual increase to a maximum with the threshold at 5 and then a steady decrease. In the previous recipe, we used a threshold of 2 as provided by the Otsu algorithm, which corresponds to the valley in between two peaks.

To understand the nature of these peaks we can display the results for the thresholds of 1, 2 and 5. We should note that a threshold of zero is the first value of the matrices, so we will display the 3D matrix at levels 2, 3 and 6.

```
figure
subplot(131)
surfdat(cellularRegion_thres(:,:,2))
subplot(132)
surfdat(cellularRegion_thres(:,:,3))
subplot(133)
surfdat(cellularRegion_thres(:,:,6))
set(gcf,'Position',[40    300    900    400]);
colormap(gray)
```

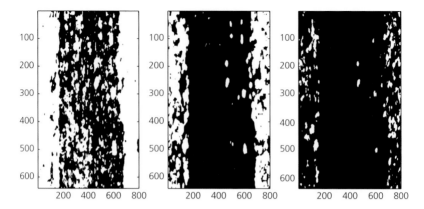

The images reveal the reason for the sudden peak at 1; the threshold at 1 is so low that it includes a good part of the background, that is, the threshold is including noise as part of the foreground. This is mostly removed with the threshold of 2, which was our previous result. With the threshold of 5 the regions with several cells have been split, together with some of the fainter cells. Increasing the threshold from here would remove cells from the segmentation.

We can repeat the overlaying on the original image to visualise the results using a higher threshold.

```
migrationImage_res              = migrationImage;
migrationImage_res(:,:,2)     = migrationImage_res(:,:,2)-...
    50*uint8(cellularRegion_thres(:,:,6));
migrationImage_res(migrationImage_res<0) = 0;
figure
imagesc(migrationImage_res)
title(strcat('Cellular Region = ',...
    num2str(100*areaCoveredbyCells(6)),...
    '[%];    Num Regions = ',...
    num2str(numRegions(6))), 'fontsize',14)
```

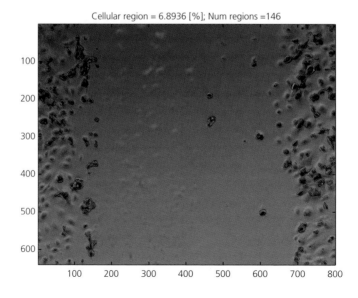

Cellular region = 6.8936 [%]; Num regions =146

One of the drawbacks of the segmentation by thresholding is the dilemma of selecting the correct threshold; a low threshold will include those regions with low intensities but may include noise. A high threshold removes noise more effectively but will then lose the faint regions.

There are several techniques to deal with situations like the one previously described. One way to avoid the problems of selecting either a low or a high threshold is to select two: a high one that detects regions that are very likely to be the signal of interest, and a low one under which all signal is most likely to be background. Those pixels whose values are in between the two thresholds are considered depending on their neighbours; if the pixels are close to a region above the high threshold, then they must be part of a region of foreground. If on the other hand, the pixels are surrounded by background, then they are also assigned to the background. This technique is called "hysteresis thresholding" or double thresholding and is used in several areas of engineering.

To calculate the segmentation by hysteresis thresholding we will use the previous threshold of 2, and to obtain the high/low thresholds we will simply add and remove 1.

```
thresLevel            = 2;
cellularRegion        = migrationImage_HPF_LPF>(thresLevel);
cellularRegion_high   = migrationImage_HPF_LPF>(thresLevel+1);
cellularRegion_low    = migrationImage_HPF_LPF>(thresLevel-1);
```

We can visualise the effects of the hysteresis thresholding in two different ways. First we will add the results of the low and high thresholds. In this way we will observe the two regions together, and in this way we can see those regions with the low threshold that are in contact with regions above the high threshold.

```
figure
imagesc((cellularRegion_high)+(cellularRegion_low))
colormap(gray)
title(['White: regions>High threshold;',...
    'Grey: regions>low threshold'],...
    'fontsize',14);
```

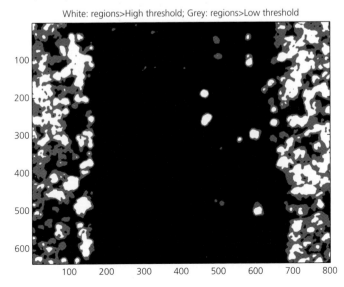

For the second display, we will remove those regions of the low threshold segmentation that are not in contact with high regions. To do that we need first to identify the regions uniquely. We can do that by labelling them with the command 'bwlabel'. Then we find which of those regions are in contact with high regions by multiplying the labelled image with the high results. Finally we discard those regions using the command 'ismember', which selects objects of a labelled image according to a certain criterion. In this case, the criterion will be to be in contact with a high region. This process is achieved with the following four commands:

```
cellularRegion_low_Labelled = bwlabel(cellularRegion_low);
overlappingRegions          = cellularRegion_low_Labelled.*...
                              cellularRegion_high;
regionsToKeep               = unique(overlappingRegions);
cellularRegion_low_final    = ismember(...
   cellularRegion_low_Labelled,regionsToKeep(2:end));
```

We can now compare the regions that we will include and remove from the original segmentation by using hysteresis thresholding.

```
figure
imagesc(cellularRegion-cellularRegion_low_final)
colormap(gray)
title('White: regions removed;    Black: regions included',...
      'fontsize',14);
```

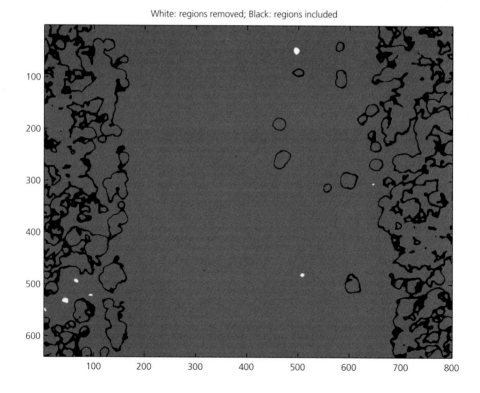

White: regions removed; Black: regions included

We can see that the majority of the pixels do not change from the single to the double thresholding, i.e. those in grey. The black regions show how the cellular regions on the edges of the image cover larger areas, and on the wound region, some regions have been removed, i.e. those in white.

Sometimes it is useful to save intermediate results that we might be using at a later stage. This is done with the command 'save', which, as the name indicates, saves to the hard drive whatever we indicate. If we type 'save' and follow that by a file name and enter, Matlab will save all the variables that have been created so far. If we want to save only a few variables, we just list the variables we want to save after the file name, for instance:

```
save results_ScratchWoundAssay_Hysteresis...
 cellularRegion_low_final migrationImage
```

would save the last result and the original image of the scratch wound assay into a file which we can later read.

5.6   Recipe 5: Morphological Operators

So far we have been identifying the regions of cells by filtering an original image with different filters, or combination of filters. We have slightly modified the resulting images by removing some of its regions, but we have not exploited all the possible modifications that we can perform to the binary images that are the output of the segmentation process. These techniques are commonly known as "morphological operations" or "mathematical morphology".

The basis of the morphological operators is a series of mathematical operations that will be performed on an image with the application of a predefined structure, sometimes called "structural elements" or "kernel". To begin our illustration of morphological operators we can load the filtering results and the original scratch wound assay we saved on the previous recipe. As the dimensions will be used later on, we can calculate the rows and columns of the segmented cellular regions at this time.

```
load results_ScratchWoundAssay_Hysteresis
[rows,columns]              = size(cellularRegion_low_final);
```

```
figure
imagesc(cellularRegion_low_final)
colormap(gray)
```

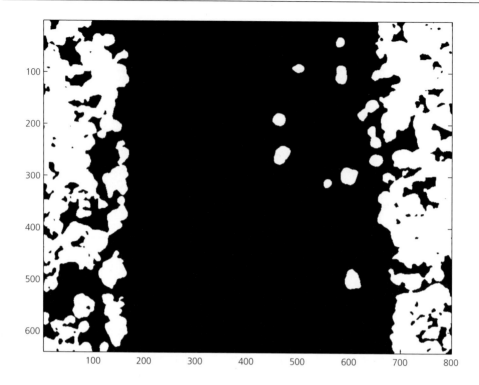

The first two basic morphological operations are used to remove or add pixels to the boundaries of all the regions of an image. The operations are called "dilation" when pixels are added and "erosion" when pixels are removed. The amount of pixels to be removed is determined by the structural element that will be used, so a small structural element will remove few pixels and vice versa. The shape of the operator is also important, as we will see in the following examples.

First we will use a small symmetrical operator formed by a [7 × 7] matrix of ones. The commands to perform dilation and erosion in Matlab are 'imdilate' and 'imerode' with the following syntax:

```
cellularDilation   = imdilate(cellularRegion_low_final,ones(7));
cellularErosion    = imerode(cellularRegion_low_final,ones(7));

figure
subplot(121)
imagesc( cellularDilation)
title('Regions after Dilation','fontsize',14);
subplot(122)
imagesc( cellularErosion)
title('Regions after Erosion','fontsize',14);
colormap(gray)
```

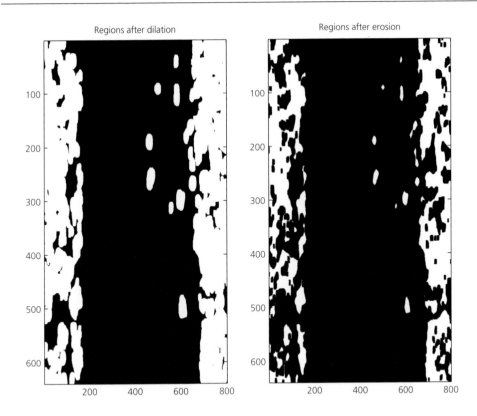

The effects of dilation and erosion are clear in the previous figure in which we can see how the dilated image covers a larger area and some of the gaps between previously unconnected regions have disappeared. On the other hand, the eroded image has split some areas, and the size of some of the isolated regions has been considerably reduced. We can infer that if we continue to erode a region, or we erode with a structural element larger than the size of the region, it is possible to remove it completely from the image.

The structural elements can have many shapes; some of the most common are squares, circles or rectangles. We can analyse the effects of dilation and erosion with a rectangle with different orientations.

```
verticalElement     =   ones(30,4);
horizontalElement   =   ones(4,30);

verticalDilation    =   imdilate(cellularRegion_low_final,...
  verticalElement);
verticalErosion     =   imerode(cellularRegion_low_final,...
  verticalElement);

horizontalDilation  =   imdilate(cellularRegion_low_final,...
  horizontalElement);
```

```
horizontalErosion   =   imerode(cellularRegion_low_final,...
  horizontalElement);


figure
subplot(121)
imagesc( verticalDilation)
title('Dilation with Vertical Element','fontsize',14);
subplot(122)
imagesc( horizontalDilation)
title('Dilation with Horizontal Element','fontsize',14);
colormap(gray)

figure
subplot(121)
imagesc( verticalErosion)
title('Erosion with Vertical Element','fontsize',14);
subplot(122)
imagesc( horizontalErosion)
title('Erosion with Horizontal Element','fontsize',14);
colormap(gray)
```

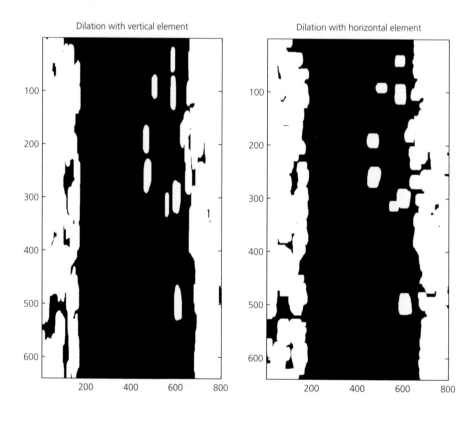

Erosion with vertical element   Erosion with horizontal element

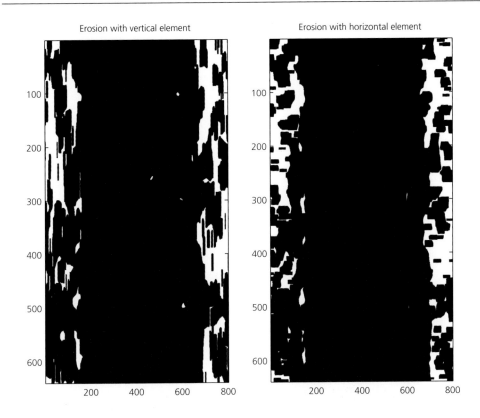

As it could be expected, the vertical structural element has stretched regions downwards and therefore has closed some regions, which were not connected before, and are not connected in the horizontal dilation. Notice the differences in the bottom left-hand side corner. The erosion presents again similar effects. Notice how in some cases there are thin lines that connect between regions. Further erosion would disconnect those regions.

Matlab provides the command 'strel' that can create structuring elements of many 2D and 3D shapes.

An interesting question arises if we consider if erosion and dilation are reversible, that is, if we erode an image and then dilate it with the same structural element, do we return to the original image? As in so many cases, it depends on the image, but generally, these operations are not reversible. Consider the case of a region that disappears after an erosion, that cannot be regenerated by a posterior dilation. Equally when a "hole" in an image is filled with a dilation, it will not be subject to erosion any more as it will not have borders to be eroded. The concatenation of erosion and reversion receives different names depending on the order in which the operations are performed: erosion followed by dilation is called "opening" and dilation followed by erosion is called "closing". The corresponding Matlab commands are 'imopen' and 'imclose'. To illustrate these operations we will use a circular morphological operator.

```
SE_Disk            =  strel('disk',10);
cellularOpening    =  imopen(cellularRegion_low_final,SE_Disk);
cellularClosing    =  imclose(cellularRegion_low_final,SE_Disk);

figure
subplot(121)
imagesc( cellularOpening+cellularRegion_low_final)
title('Opening with Disk Structural Element','fontsize',14);
subplot(122)
imagesc( cellularClosing+cellularRegion_low_final)
title('Closing with Disk Structural Element','fontsize',14);
colormap(gray)
```

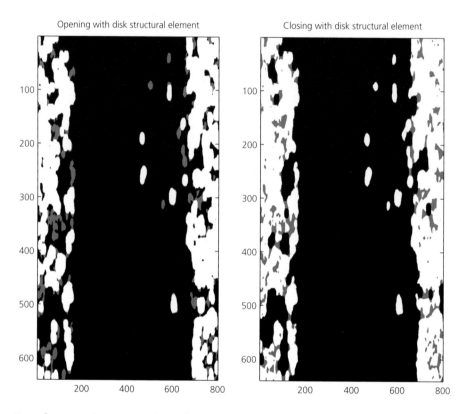

For the opening operation, the regions in white correspond to result of the operation, and the regions in grey correspond the areas of the original image that have been removed or "opened". For the closing, the areas in grey are those areas not on the original image that have been "closed".

Now that we have seen the effects of a few morphological operators, the possibilities of processing binary images are very attractive as it is possible to remove small elements due to noise, connect empty regions, obtain skeletons from images, detect corners or bridging pixels, etc. For example we can fill in holes inside images using the command 'imfill'.

```
cellularRegion_noHoles    =    (imfill(cellularClosing,'holes'));

figure
imagesc( cellularRegion_noHoles+cellularClosing)
colormap(gray)
title('Image with holes filled','fontsize',14);
```

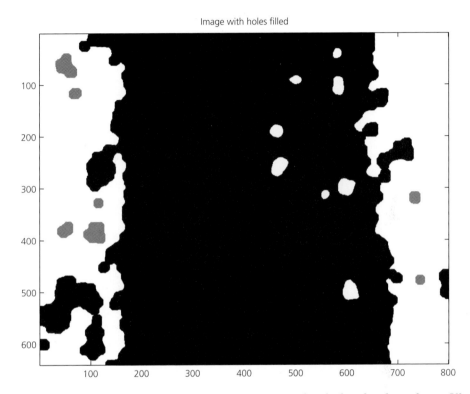

In the previous image the grey regions correspond to holes that have been filled. If we consider that for a scratch wound assay we are interested in detecting the two main regions of cells on the edges of the image and the large region in between as the wound area, we then can play with the dimensions of the structural elements to obtain the desired segmentation. For this effect, we can employ larger structural elements. Firstly, we will close all the gaps in the cells so that there are no holes or other protrusions (like "peninsulas" from the "main land" if we think of the cellular regions as land and the wound or background as water). Secondly, we will remove any smaller regions in the wound area. We can try for instance to have structural elements of sizes 30 and 20 with shape of a disk to do the closing and opening, respectively.

```
SE_Disk30            =    strel('disk',30);
SE_Disk20            =    strel('disk',20);

cellularRegion_1     =    (imclose(cellularRegion_low_final,...
   SE_Disk30));
```

```
cellularRegion_2    =   (imopen(cellularRegion_1>0,SE_Disk20));


figure
imagesc(cellularRegion_2)
title('Regions after Structural Closing (larger elements)',...
    'fontsize',14);
colormap(gray)
```

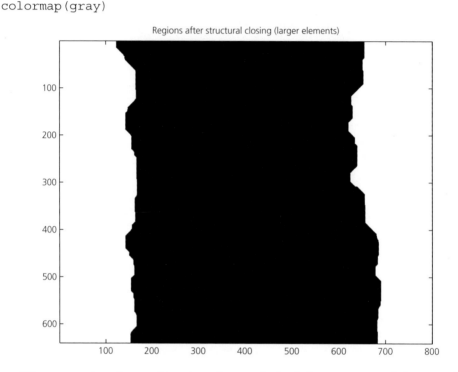

We can see that the combination of two relatively large structural elements have successfully removed all small elements in the area of the wound and have consolidated the two regions at the edges as solid regions. We can overlay the results on the original image and calculate the area covered by the cells. As a reference we had previously obtained 17.85% with the combination of low pass and high pass filters.

```
areaCoveredByCells      =   sum(cellularRegion_2(:))/rows/...
    columns;


figure
migrationImage_res      =   migrationImage;
migrationImage_res(:,:,2)   = migrationImage_res(:,:,2)-...
    50*uint8(cellularRegion_2);
migrationImage_res(migrationImage_res<0) = 0;
figure
imagesc(migrationImage_res)
title(strcat('Cellular Region = ',...
    num2str(100*areaCoveredByCells),' [%];'),'fontsize',14)
```

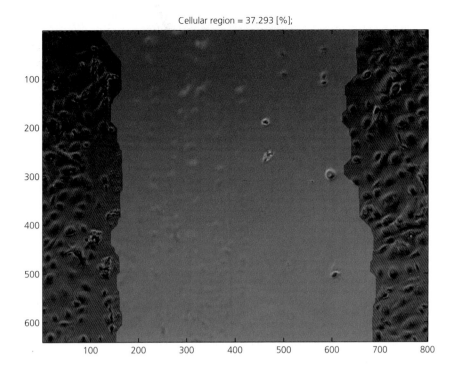

Cellular region = 37.293 [%];

The results show how the two cellular areas uniformly cover regions with cells even where there are some gaps in between them and the wound region is now excluding small isolated cells. The area covered by cells has increased to 37.29%.

There are many other important morphological operations that can be useful; several of these can be obtained with a single Matlab command, 'bwmorph', which stands for binary morphological operation. The command takes as input arguments the image to be modified and the specific operation. In some cases a third parameter may be used. We will illustrate just a few interesting cases.

In some cases, we might be interested to measure not only the area, but the perimeter of a certain region as a compact region could have the same area as an elongated one, yet their perimeters could be very different. One way of obtaining the perimeter could be to remove all the internal pixels of the regions and leaving only the ones in the border between the background and foreground. This is obtained by selecting the operation 'remove' of 'bwmorph'. To better illustrate the properties of the algorithms we will use the results after obtained after thresholding the filtered image.

```
cellularBoundary           =   bwmorph(cellularRegion_low_final,...
                               'remove');


figure
imagesc(imdilate(cellularBoundary,ones(3)))
title(strcat('Boundary Cellular Region = ',...
    num2str(sum(cellularBoundary(:)))),'fontsize',14)
colormap(gray)
```

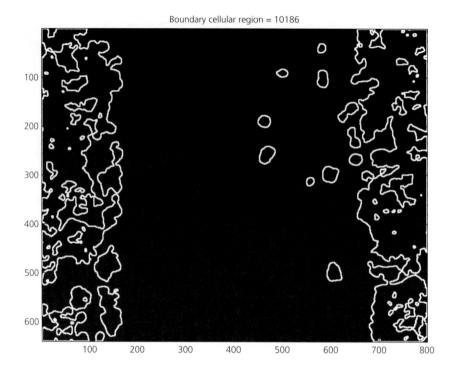

Boundary cellular region = 10186

Notice that for display purposes we have dilated the results so that the lines of the figure appear thicker.

In other cases we might be interested to find the central line of a region, for instance with cells that are elongated, like neurons, we might be interested in finding the centre of an axon. We can obtain the central lines by reducing the objects to a line of width of one pixel that follows the shapes of the objects in several ways; each algorithm has its own advantages and disadvantages depending on the context. We will illustrate two very common ones: "skeletonising" (also known as medial axis transform) and "thinning". To better illustrate the properties of the algorithms we will use the results after opening and closing with large structural elements.

```
cellularRegionSkeletonised = bwmorph(cellularRegion_2,'skel',...
                             'inf');
cellularRegionThinned      = bwmorph(cellularRegion_2,'thin',...
                             'inf');
```

```
figure
imagesc(cellularRegion_2+imdilate(cellularRegionSkeletonised,...
       ones(3)))
colormap(gray)
```

```
figure
imagesc(cellularRegion_2+imdilate(cellularRegionThinned,...
       ones(3)))
colormap(gray)
```

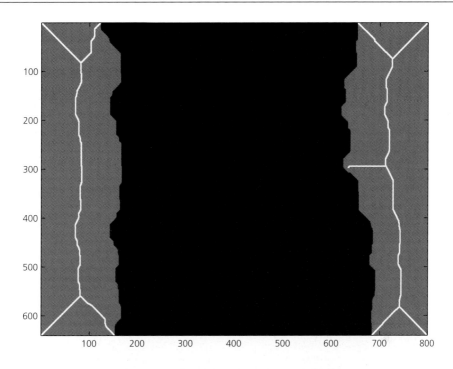

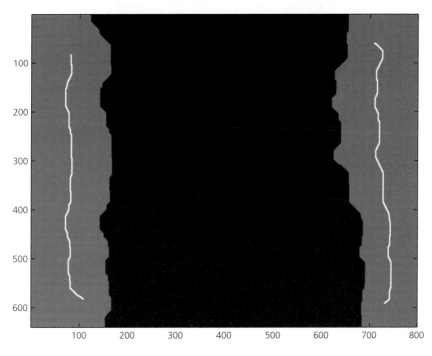

Both skeletonising and thinning have reduced the regions into 1-pixel wide lines. Whilst thinning has resulted in a line with no branches, skeletonising has produced several interconnected segments. In fact, skeletonising is very sensitive to small variations of the shape, whilst thinning produces a more "central" line. On the other hand, if we are interested in the shapes through branching points, then skeletonising may provide better results. We can also obtain "end points" and "branch points" with bwmorph in the following way:

```
endPointsSkeleton       =   bwmorph(cellularRegionSkeletonised,...
                            'endpoints');
branchPointsSkeleton    =   bwmorph(cellularRegionSkeletonised,...
                            'branchpoints');
numEndPoints            =   sum(endPointsSkeleton(:));
numBranchPoints         =   sum(branchPointsSkeleton(:));

figure
imagesc(imdilate(cellularRegionSkeletonised,ones(3))+...
    2*imdilate(endPointsSkeleton,SE_Disk)+...
    3*imdilate(branchPointsSkeleton,SE_Disk))
title(strcat('Branch points = ',num2str(numBranchPoints),...
    ';   End Points = ',num2str(numEndPoints)),'fontsize',14)
```

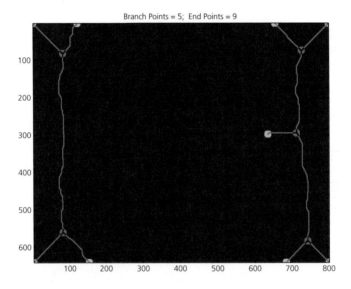

As we have done previously, we will save our current results, which will become useful in the future.

```
save results_ScratchWoundAssay_Morphology ...
    migrationImage cellularRegion_low_final ...
    cellularRegion_2 rows columns areaCoveredByCells
```

5.7   Recipe 6: Measuring Distances Between Cellular Boundaries

There is one final measurement that we would like to obtain: the distance between the two regions of cells. At the beginning of this chapter we mentioned that it is a common practice to draw several lines between one boundary of the cells with the background towards the other boundary. These values are then averaged to obtain a more reliable measurement, as the boundaries are rarely straight lines. In this recipe we will investigate one technique to measure distances. First, we will retrieve the results we obtained in the previous recipe as we are interested in measuring the boundaries of the scratch wound assay that was processed with the large morphological operators and the left two cellular regions on the edges and one wound area in the centre.

```
load results_ScratchWoundAssay_Morphology
```

First of all we would like to locate and distinguish the two boundaries of the cellular regions with the background. That is, we could locate the boundaries with the morphological operator 'bwmorph' with 'remove', however, that would not distinguish the boundaries directly. An easy way to distinguish the boundaries is by subtracting the horizontal neighbouring pixels. This operation will take advantage of the fact that the boundary on the left side of the wound drops from 1 to 0 whilst the boundary on the right raises from 0 to 1. The next figure illustrates this by displaying a profile line of the last row of the image of the segmented scratch wound assay.

```
figure
subplot('Position', [0.10     0.35     0.85     0.60])
imagesc(cellularRegion_2)
subplot('Position', [0.10     0.10     0.85     0.20])
plot(cellularRegion_2(rows,:),'linewidth',2)
axis ([-1 columns+1 -0.05 1.05])
ylabel('Profile','fontsize',14)
xlabel('Column Position','fontsize',14)
grid on
colormap(gray)
```

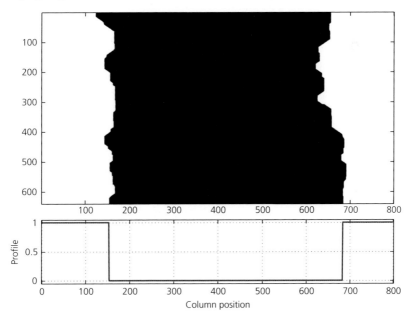

Notice that we have created two plots of different dimensions in the same figure, without using the predefined sizes that are generated with the normal options of 'subplot', like 'subplot (3,2,1)', which partitions the figure in three rows, two columns and starts at the top left, i.e. subplot number 1. We can create subplot or 'axes' on the figure in any position that we specify with the coordinates in relative terms to the figure (i.e. between 0 and 1). The position is defined with four parameters in the following order [initial-column initial-row size-columns size-rows].

We can now calculate the difference between pixels in the horizontal dimension in two different ways. First, we can use two copies of the matrix, but shifting them by one column like this:

```
edgesRegion = [cellularRegion_2(:,2:end)- ...
               cellularRegion_2(:,1:end-1) zeros(rows,1)];
```

Notice that since we will shift the image by one column, we need to append a column of zeros so that the final result has the same dimensions as our original image. The decision to append the column of zeros at the right side of the image is arbitrary and could also be pre-pended on the left or somewhere in between.

Alternatively, we can use the command 'diff', which finds the differences between elements of a matrix. We also can indicate how many times we want to obtain the differences and in which dimension. For our example, we type:

```
edgesRegion = [diff(cellularRegion_2,1,2) zeros(rows,1)];
```

We can display the results in the same way as before, the image and one profile.

```
figure
subplot('Position', [0.10     0.35     0.85     0.60])
imagesc(edgesRegion)
subplot('Position', [0.10     0.10     0.85     0.20])
plot(edgesRegion(rows,:),'linewidth',2)
axis ([-1 columns+1 -1.05 1.05])
ylabel('Profile Edges','fontsize',14)
xlabel('Column Position','fontsize',14)
grid on
colormap(gray)
```

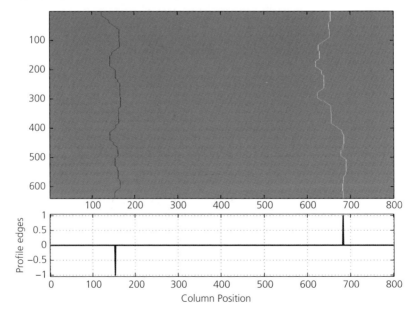

One of the edges is now positive whilst the other is negative, so we can easily identify them with a comparison with zero. The next step involves one very useful Matlab command, 'bwdist', which calculates distances on binary images. The command 'bwdist' calculates the distance, which may be defined in many ways, from every pixel in the image, to its nearest non-zero pixel. Therefore we can calculate the distances of the pixels on the image based on each of the boundaries previously determined and display.

```
distanceFromLeft        =   bwdist(edgesRegion<0);
distanceFromRight       =   bwdist(edgesRegion>0);

figure
subplot('Position', [0.10    0.35    0.85    0.60])
imagesc(distanceFromLeft)
colorbar
subplot('Position', [0.10    0.10    0.85    0.20])
plot(distanceFromLeft(rows,:),'linewidth',2)
ylabel('Distance','fontsize',14)
xlabel('Column Position','fontsize',14)
axis tight
grid on
colormap(gray)
```

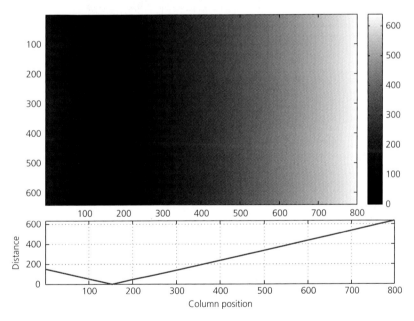

The distances have been calculated for every pixel and appear as shades of grey. We now need to find out the intersection of each boundary on that distance map generated from the opposite boundary. In this way we will find the minimum distances of each point in the boundary towards the opposite boundary. We can visualise this by removing each boundary from the distance maps.

```
figure
subplot('Position', [0.10    0.30    0.85    0.30])
imagesc(distanceFromLeft.*(1-imdilate(edgesRegion>0,ones(5))));
```

```
title('Distances from Left Boundary','fontsize',14);
subplot('Position', [0.10    0.65    0.85    0.30])
imagesc(distanceFromRight.*(1-imdilate(edgesRegion<0,...
        ones(5))));
title('Distances from Right Boundary','fontsize',14);

subplot('Position', [0.10    0.05    0.85    0.20])
hold on
plot(distanceFromLeft(rows,:),'b-','linewidth',2)
plot(distanceFromRight(rows,:),'r--','linewidth',2)
plot(510*abs(edgesRegion(rows,:)),'k-','linewidth',2)

title('Distances from Edges','fontsize',14);
axis tight
grid on
set(gcf,'Position',[40    200    600    700]);
colormap(gray)
```

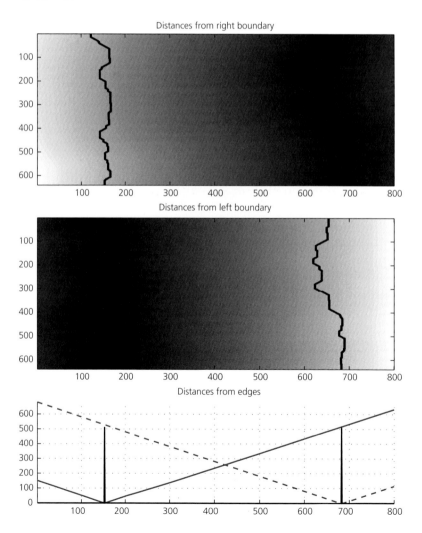

For visualisation purposes we have dilated the boundaries so that they are clearly visible. The bottom plot shows the exact values of distance for the last row of the image; in this case, the distances are around 500 pixels in distance. Notice that the distances are not exactly the same; they depend on the shape of each boundary.

To calculate the distances and find the minimum, maximum and average distance values in three steps, first we find the intersection of the boundaries and the distance maps. Next we identify all unique values of the intersection. Finally we discard the zeros (it should be the first value of the unique list of values) and use the commands 'min', 'mean' and 'max' to find our three values of interest.

```
distanceAtLeftBoundary       = distanceFromRight.*...
    (edgesRegion<0);
distanceAtRightBoundary      = distanceFromLeft.*...
    (edgesRegion>0);

distanceAtLeftBoundary_list = unique(distanceAtLeftBoundary);
distanceAtRightBoundary_list= unique(distanceAtRightBoundary);

distanceBetweenBoundariesLeft   = floor([...
    min(distanceAtLeftBoundary_list(2:end))...
    mean(distanceAtLeftBoundary_list(2:end))...
    max(distanceAtLeftBoundary_list(2:end))]);

distanceBetweenBoundariesRight  = floor([...
    min(distanceAtRightBoundary_list(2:end))...
    mean(distanceAtRightBoundary_list(2:end)) ...
    max(distanceAtRightBoundary_list(2:end))]);

disp([distanceBetweenBoundariesLeft;...
    distanceBetweenBoundariesRight])


    457    486    529
    457    492    529
```

We can see that the minimum and maximum distances are the same whether we calculate from the left or from the right. The average distance is slightly different. To avoid any bias, we can define the average distance as the mean of these two values:

```
distanceBetweenBoundaries   = ...
    (distanceBetweenBoundariesLeft(2)+...
    distanceBetweenBoundariesRight(2))/2;
```

Finally we can display the results together in a single figure. To illustrate how easy it is to modify the output we can calculate and shade the area of the wound instead of the cellular regions by subtracting the value of the cells and the cellular regions from the value of 1.

```
migrationImage_res2        = migrationImage;
migrationImage_res2(:,:,2) = migrationImage_res2(:,:,2)+...
```

```
    50*uint8(1-cellularRegion_2);
migrationImage_res2(migrationImage_res2>255) = 255;
figure
imagesc(migrationImage_res2)
title(strcat('Wound Area = ',...
    num2str(1-areaCoveredByCells,2),...
    '[%];  Av. Distance = ',...
    num2str(distanceBetweenBoundaries),...
    '[pix]'),'fontsize',14)
```

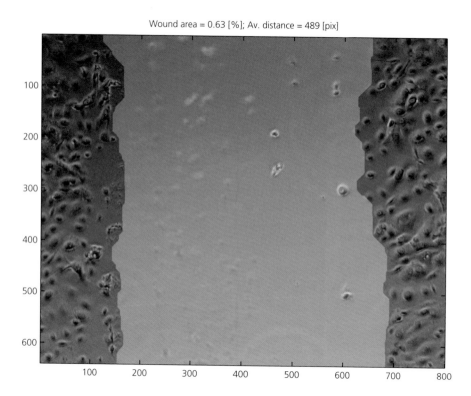

Wound area = 0.63 [%]; Av. distance = 489 [pix]

To finish this recipe we will repeat the segmentation process on an image of a scratch wound assay with different characteristics. We will first apply the combination of low and high pass filters and threshold, and later apply morphological operators to filter the image of the regions of small detail.

```
migrationImage2 = imread(...
  'http://caiman.group.shef.ac.uk/caiman/images/migration5.jpg');
figure
imagesc(migrationImage2)
```

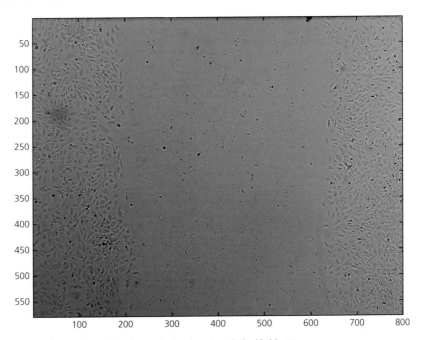

Image courtesy of Dr. Chryso Kanthou, The University of Sheffield, UK.

```
lowPassFilter_2         = fspecial('Gaussian', [15 15],3);

migrationImage2_LPF     = imfilter(migrationImage2(:,:,1),...
                            lowPassFilter_2,'replicate');
migrationImage2_HPF     = uint8(double(...
                        migrationImage2(:,:,1))-...
                        double(migrationImage2_LPF));
migrationImage2_HPF_LPF = imfilter(abs(migrationImage2_HPF),...
                            lowPassFilter_2);



thresLevel2             = 255*graythresh...
   ((migrationImage2_HPF_LPF));
cellularRegion2         = ...
   migrationImage2_HPF_LPF>=(thresLevel2-1);



migrationImage2_res     = migrationImage2;
migrationImage2_res(:,:,2)  = migrationImage2_res(:,:,2)-...
                        100*uint8(cellularRegion2);
migrationImage2_res(migrationImage2_res>255) = 255;
migrationImage2_res(migrationImage2_res<0) = 0;
```

```
figure
imagesc(migrationImage2_res)
title(strcat('Cellular Area = ',...
    num2str(sum(cellularRegion2(:))/rows/columns,3),...
    '[%]'),'fontsize',14)

SE05                      =   strel('disk',10);
SE06                      =   strel('disk',30);
cellularRegion2_A         =   (imopen(cellularRegion2,SE05));
cellularRegion2_B         =   (imclose(cellularRegion2_A>0,...
    SE06));

figure
migrationImage3_res              = migrationImage2;
migrationImage3_res(:,:,2)   = migrationImage3_res(:,:,2)-...
    100*uint8(1-cellularRegion2_B);
migrationImage3_res(migrationImage3_res>255) = 255;
migrationImage3_res(migrationImage3_res<0) = 0;

imagesc(migrationImage3_res)
title(strcat('Wound Area = ',...
    num2str(1-sum(cellularRegion2_B(:))/rows/columns,3),...
    '[%]'),'fontsize',14)
```

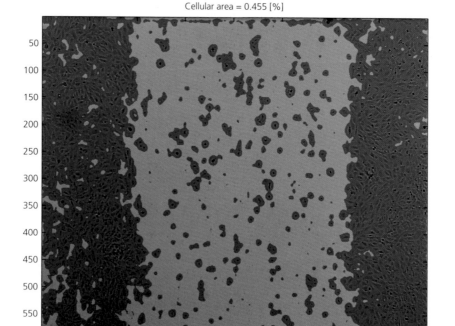
Cellular area = 0.455 [%]

Wound area = 0.597 [%]

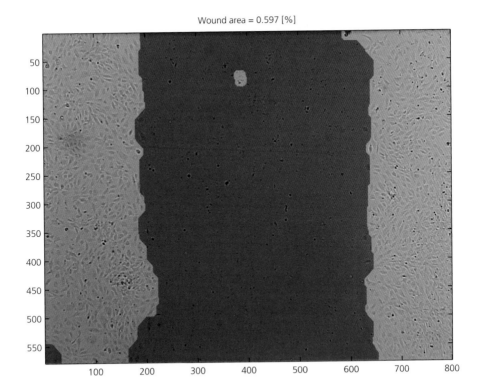

We can see that the analysis process is consistent even when the input images present different characteristics.

5.8   Recipe 7: Introduction to Fourier Analysis

A very important topic in image analysis is based on a mathematical operation called "Fourier Transform", which was developed by the French Mathematician Joseph Fourier. The essence of the Fourier transform is a change of the space or domain of analysis of a function. We normally think of signals in one or more dimensions (like electrocardiograms in 1D or images in 2D) as a series of values that change with time (ECGs) or space (images). These signals are considered to be in the "time-domain" or "spatial-domain" as the variation of the intensities follows their position in time or in space. The Fourier transform will translate those signals into a different domain in which the variation of the intensities depend on the frequency of the signal, and therefore these are signals in the Frequency-domain, which is sometimes referred to as the Fourier-domain. Mathematically, the basis of Fourier transform is the analysis of exponential Fourier series, that is, to represent a signal by the sum of the exponential signals that are orthogonal to each other. The exponential signals of the Fourier transform are formed by sine and cosine functions; the sine function is multiplied by the square root of -1, which implies that the function is of complex nature.

Rather than writing down the equations, which may be confusing, we will define the Fourier transform with the following example.

We will start by analysing a simple square wave. This signal is easily defined by creating a sine function that oscillates at a certain frequency and then taking the sign (negative, positive or zero) of that sinusoidal signal. We can create those signals in the following way:

```
spaceAxis            = 8*(0:255)/256;
SinusoidWave_1       = (sin(1*2*pi*spaceAxis));
SquareWave           = sign(SinusoidWave_1);
```

Notice that we created one extra signal for the spatial axis. This signal will be useful later on. We defined the axis as 256 points that go from 0 to 8 so that we can have 8 cycles in the Sinusoid and the square waves.

We will now display these two signals together, and, since we are interested in comparing them, we will scale the square wave by a factor of 0.8. This scaling will reduce the difference between the two waves and we can also plot that difference.

```
figure
subplot(211)
plot(spaceAxis,0.8*SquareWave,'b-',...
    spaceAxis,SinusoidWave_1,'r--','linewidth',2)
axis tight
grid on
title(['Square (solid blue) and Sinusoidal',...
    '(dashed red) Waves'],'fontsize',14);
xlabel('Space Axis','fontsize',14)
ylabel('Amplitude','fontsize',14)
subplot(212)
plot(spaceAxis,0.8*SquareWave -SinusoidWave_1,'b-',...
    'linewidth',2)
axis tight
grid on
title('Difference between Square  and Sinusoidal  Waves',...
    'fontsize',14);
xlabel('Space Axis','fontsize',14)
ylabel('Amplitude','fontsize',14)
subplot(212)
```

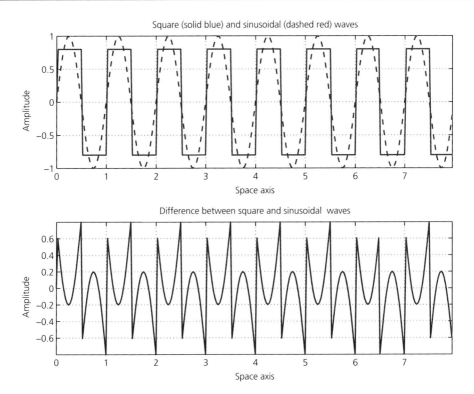

It is obvious that the two signals are very different. One way to reduce the difference between the two waves is to add a second signal to the Sinusoidal, for instance, if we generate a new sinusoidal with the frequency multiplied by 3 and the amplitude reduced by 3. The following image compares the sum of the signals against the square wave and the two signals together.

```
SinusoidWave_3        = (sin(3*2*pi*spaceAxis))/3;

figure
subplot(211)
plot(spaceAxis,0.8*SquareWave,'b-',...
    spaceAxis,SinusoidWave_1+SinusoidWave_3,'r--',
    'linewidth',2)
axis tight
grid on
title('Approx. of a Square Wave with 2 Sinusoidal Waves',...
    'fontsize',14);
xlabel('Space Axis','fontsize',14)
ylabel('Amplitude','fontsize',14)

subplot(212)
plot(spaceAxis,SinusoidWave_1,'b-.',...
    spaceAxis,SinusoidWave_3,'k-','linewidth',2)
axis tight
```

```
grid on
title(['Two Sinusoidal Waves of different amplitudes',...
    'and frequencies'],'fontsize',14);
xlabel('Space Axis','fontsize',14)
ylabel('Amplitude','fontsize',14)
```

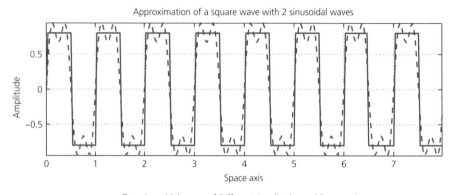

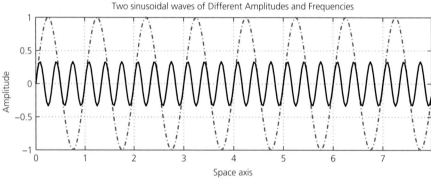

From the top graph it is evident that the Square wave is now closer to the dashed line created by the sum of the signals. We can observe that these two signals share some properties, they both cross 0 at the same positions for instance.

We can continue the process of adding signals; for instance we can add three more with higher frequencies and lower intensities in the following way:

```
SinusoidWave_5        = (sin(5*2*pi*spaceAxis))/5;
SinusoidWave_7        = (sin(7*2*pi*spaceAxis))/7;
SinusoidWave_9        = (sin(9*2*pi*spaceAxis))/9;

figure
subplot(211)
plot(spaceAxis,0.8*SquareWave,'b-',...
    spaceAxis,SinusoidWave_1+SinusoidWave_3+...
    SinusoidWave_5+SinusoidWave_7+SinusoidWave_9,'r--',...
    'linewidth',2)
axis tight
grid on
```

```
title('Approx. of a Square Wave with 5 Sinusoidal Waves',...
    'fontsize',14);
xlabel('Space Axis','fontsize',14)
ylabel('Amplitude','fontsize',14)


subplot(212)
plot(spaceAxis,SinusoidWave_1,'b--',...
    spaceAxis,SinusoidWave_3,'k-',...
    spaceAxis,SinusoidWave_5,'m--',...
    spaceAxis,SinusoidWave_7,'r:',...
    spaceAxis,SinusoidWave_9,'b-','linewidth',2)
axis tight
grid on
title(['Five Sinusoidal Waves of different amplitudes',...
    'and frequencies'],'fontsize',14);
xlabel('Space Axis','fontsize',14)
ylabel('Amplitude','fontsize',14)
```

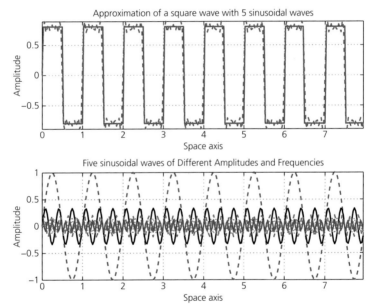

The combination of the five sinusoidal signals is now very close to the square wave. In theory, if we continue adding signals with the same pattern (increasing frequency by a factor of n+2 a reducing amplitude by the same factor of n+2), we could create a perfect square wave once we add an infinite number of sinusoids.

We can notice from the bottom plot of the previous figure that whilst all the sinusoids are displayed on the same axis, we could also re-arrange them in a different way by exploiting the fact that they have different frequency characteristics.

For this purpose, we can display them with the same axes of amplitude and space, but we will create a new axis, one that corresponds to the frequency of each sinusoid. We can do that by multiplying two matrices, one with values from 1 to 9 (the highest frequency we used) and one that spans from 1 to 256 (the number of points of our signals).

```
frequencyPos            = (1:9)'*ones(1,256);
```

We can now plot the signals in a three-dimensional graph each at the corresponding frequency position:

```
figure
hold on
plot3(frequencyPos(1,:),spaceAxis,SinusoidWave_1,'b-.',...
      'linewidth',3)
plot3(frequencyPos(3,:),spaceAxis,SinusoidWave_3,'k-',...
      'linewidth',3)
plot3(frequencyPos(5,:),spaceAxis,SinusoidWave_5,'m--',...
      'linewidth',3)
plot3(frequencyPos(7,:),spaceAxis,SinusoidWave_7,'r:',...
      'linewidth',2)
plot3(frequencyPos(9,:),spaceAxis,SinusoidWave_9,'b-',...
      'linewidth',2)

view(60,20)

axis([ 0 9 0 8 -1 1])
grid on
title ('Five Sinusoidal Waves','fontsize',14);
xlabel('Frequency Axis','fontsize',14)
ylabel('Space Axis','fontsize',14)
zlabel('Amplitude','fontsize',14)
```

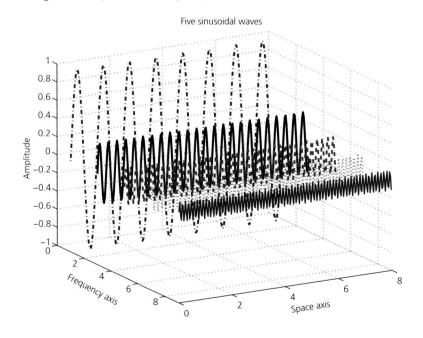

We can now see the five signals each in a separate position in the frequency axis. We can rotate the graph to give more emphasis to the frequency axis:

`view(20,10)`

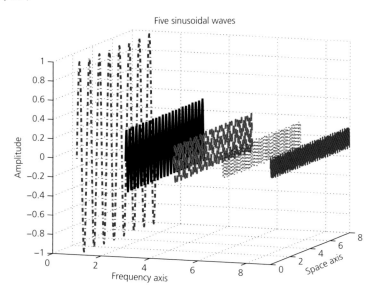

And we can rotate it even more until we no longer observe the variation of the signals over the space axis; we only see the variation over the frequency axis.

`view(0,0)`

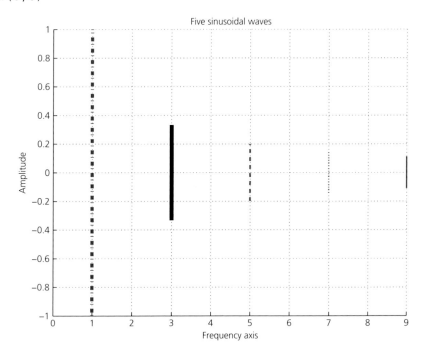

The previous process, one in which we disregard the variations of a signal over a space or time axis, to concentrate over the variations of the same signal over the frequency axis is exactly what a Fourier transform does. It converts a signal from the space or time domain into the frequency domain.

The mathematical calculation of the Fourier transform is a computationally intensive process. However, Matlab calculates the Fourier transform through a fast algorithm developed by Cooley and Tukey. This algorithm is commonly known as the "Fast Fourier Transform", and it revolutionised the field and allowed the widespread use of these transforms.

As was mentioned previously, the Fourier transform involves signals that are complex, that is, they include the imaginary number 'i' (or square root of −1) and thus they have a real part and an imaginary part, or conversely we can analyse their magnitude and phase. For the time being we will concentrate on the magnitude of the signal, which can be calculated with the command 'abs' for absolute value. A second implication of the complex nature of the transform is that the Fourier transform will have a symmetrical nature, with half of the points of the signal corresponding to positive frequencies and half to negative frequencies. For our 1D signal of this example we will only observe the positive frequencies; for images we will observe both.

The command for the Fast Fourier Transform in Matlab is 'fft' for 1D signals, 'fft2' for 2D signals and 'fftn' for higher dimensional signals. So, to calculate the Fourier transform of our square wave we do it in the following way:

```
SquareWave_frequency          = abs((fft(0.8*SquareWave)))/128;
```

The signal has been normalised with two variables, 0.8, which we had used previously to compare the Square wave with the sinusoids, and 128, which corresponds to half the number of points of the signal. That is, we will only plot half of the values of the signal.

To display the Fourier transform, we also require a frequency axis, which can be derived from the space axis with adequate normalisation (we had 8 cycles).

```
frequencyAxis                 = 8/2*spaceAxis;
```

Now we can display the Fourier transform of the square wave:

```
figure
plot(frequencyAxis(1:128),SquareWave_frequency(1:128))
axis tight
title('Fourier Transform of a Square Wave','fontsize',14);
xlabel('Frequency Axis','fontsize',14)
ylabel('Amplitude','fontsize',14)
```

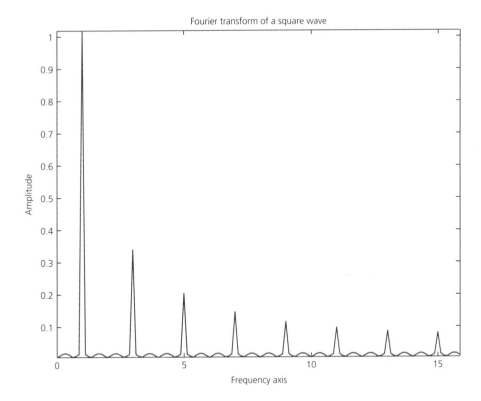

We can see that the transform has values at the frequencies 1, 3, 5, 7 etc. as we have observed previously. The magnitudes of the signals at those frequencies decrease in function of the frequency. These values are sometimes called "spikes". We can also notice that there are small values in between these frequencies, which are due to the way the transform is calculated. We can now see what the Fourier transform of a 1D signal looks like. This is sometimes called the "Fourier spectrum" of a signal, and studying signals in the Fourier domain is sometimes called "Spectral Analysis".

We could calculate transforms of other signals, like a line profile of one of the scratch wound assays, but we will first extend the Fourier transform to two dimensions.

The image of the first example has the same characteristics as the square wave, e.g. the intensities of the pixels alternate between high and low levels in one of its dimensions. It is very easy to create such an image from our square wave. We only need to do a matrix multiplication of the square wave against a vertical matrix of ones:

```
SquareField_1 = ones(256,1)*SquareWave;
```

Next, we calculate the 2D Fourier transform:

```
SquareField_1_Frequency = abs(fftshift(fft2(SquareField_1)));
```

Notice that in addition of the absolute values, we used the command 'fftshift'. This command rearranges the Fourier transform to place frequency

0 at the centre, i.e. for a 1D signal, this would put the negative frequencies (which we ignored in the previous plot) on the left and the positive frequencies (the ones we plotted) on the right. We can now display the 2D signal and its Fourier transform as follows:

```
figure
subplot(121)
imagesc(SquareField_1)
title('Two-dimensional Square Wave (H)','fontsize',12);
subplot(122)
imagesc(SquareField_1_Frequency)
title('Fourier Transform of a 2D Square Wave (H)',...
    'fontsize',12);
colormap(gray)
set(gcf,'Position',[40    300    900    400]);
```

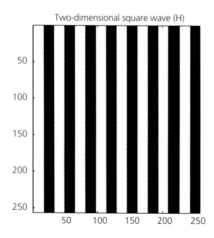

The previous figure requires careful observation, as it could seem that the transform would be a black figure, i.e. all zeros. However, in the middle of the image, there is a sequence of dots of different intensities. These dots are the equivalent of the 1D spikes. Some signals will have some frequency components, that is the spikes, that can be of a very high intensity compared to their neighbouring elements. If we zoom into the image, these are easier to visualise:

```
axis ([50 200 127 131 ])
```

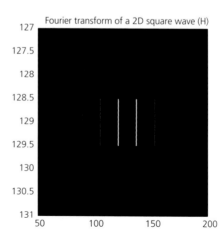

A second important observation is the symmetry of the spikes, e.g. there are two equally bright at the centre and then they decrease in intensity towards the right and the left hand side in a symmetrical way.

We can now extend this example in two ways. First we will transpose the image, that is, change the rows for columns and vice versa so that the variation of the intensity is now over the rows instead of the columns.

```
SquareField_2 = SquareField_1';
```

We can now obtain the Fourier transform and observe that the dots have also changed orientation from the horizontal axis to the vertical axis.

```
SquareField_2_Frequency = abs(fftshift(fft2(SquareField_2)));
figure
subplot(121)
imagesc(SquareField_2)
title('Two-dimensional Square Wave (V)','fontsize',12);
subplot(122)
imagesc(SquareField_2_Frequency)
title('Fourier Transform of a 2D Square Wave (V)',...
      'fontsize',12);
colormap(gray)
set(gcf,'Position',[40    300    900    400]);
```

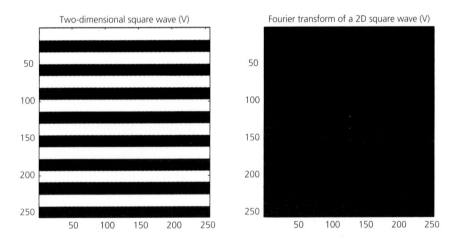

We can also combine the two square wave images to create a checkerboard pattern. This board will have variations on both dimensions, so we can expect to have dots in both axes. To highlight these dots, we will display the logarithm of the values of the Fourier transform. The effect of the logarithm is that it will reduce very high values by a very large effect whilst reducing lower values by a lower effect. For example, the logarithm base 10 ($\log_{10}$) of 10 is 1, of 100 is 2 and of 1000 is 3.

```
SquareField_3    = SquareField_1+SquareField_2;
SquareField_3_Frequency = (fftshift(fft2(SquareField_3)));
figure

subplot(121)
imagesc(SquareField_3)
```

```
title('Two-dimensional Checkerboard','fontsize',12);
subplot(122)
imagesc(log10(abs(SquareField_3_Frequency)))
title(['Fourier Transform of a 2D Checkerboard',...
    '(log of amplitude)'],'fontsize',12);
colormap(gray)

set(gcf,'Position',[40    300    900    400]);
```

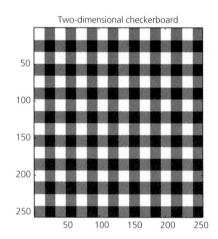
Two-dimensional checkerboard

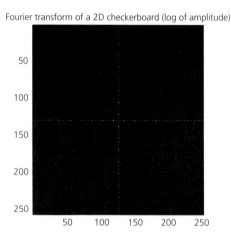
Fourier transform of a 2D checkerboard (log of amplitude)

We can now see not only the spikes due to the frequencies of the square wave as two faint lines in the horizontal and vertical at the centre of the image. Notice again that there is a plane of symmetry to the lines and the spikes.

So far we have looked at the conversion from the spatial domain into the frequency domain. The Fourier transform is reversible, and we can convert from the frequency domain back to the spatial or time domain with the commands 'ifft', 'ifft2' and 'ifftn' for 1, 2, and n-dimensions, respectively. The inverse Fourier transform is also a complex operation so we have to use the magnitude and phase of the numbers.

To extend the example of the inverse transform, we can modify one of the signals in the frequency domain and see how that impacts in the spatial domain. For the example, we will add two spikes in the diagonal:

```
SquareField_4_Frequency              =    SquareField_3_Frequency;
SquareField_4_Frequency(105,105) =    (SquareField_3_Frequency...
                                        (129,105));
SquareField_4_Frequency(153,153) = -(SquareField_3_Frequency...
                                        (129,105));
```

We copied the value of one of the spikes on the upper right quadrant and then changed the sign for the lower right quadrant. We can now display the inverse signal, again calculating the magnitude.

```
figure
subplot(121)
```

```
imagesc(abs(ifft2(SquareField_4_Frequency)))
title('Two-dimensional Partial Checkerboard','fontsize',12);
subplot(122)
imagesc(log10(abs(SquareField_4_Frequency)))
title(['Fourier Transform of a 2D Partial Checkerboard',...
    '(log of amplitude)'],'fontsize',12);
colormap(gray)

set(gcf,'Position',[40    300    900    400]);
```

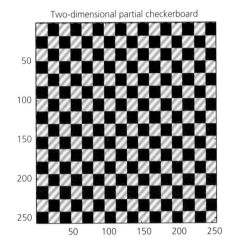
Two-dimensional partial checkerboard

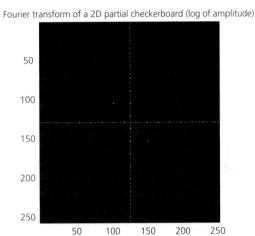
Fourier transform of a 2D partial checkerboard (log of amplitude)

We can now observe that the spikes in the diagonal have resulted in a sinusoidal variation on the image, in a diagonal.

Not all variations of the Fourier transforms are as nice and regular as square or sinusoidal waves or spikes. We can observe this by making very slight variations to the checkerboard pattern. We will set to 0 a small region of the checkerboard and observe how it impacts in the Fourier domain.

```
SquareField_6              = SquareField_3;
SquareField_7              = SquareField_3;

SquareField_6(1:16,1:16)   = 0;
SquareField_6_Frequency    = abs(fftshift(fft2(SquareField_6)));

SquareField_7(1:5,1:100)   = 0;
SquareField_7_Frequency    = abs(fftshift(fft2(SquareField_7)));


figure
subplot(121)
imagesc(SquareField_6)
title('Two-dimensional Partial Checkerboard','fontsize',12);
subplot(122)
imagesc(log10(SquareField_6_Frequency))
```

```
title(['Fourier Transform of a 2D Partial Checkerboard',...
    '(log of amplitude)'],'fontsize',12);
colormap(gray)

set(gcf,'Position',[40    300    900    400]);


figure
subplot(121)
imagesc(SquareField_7)
title('Two-dimensional Partial Checkerboard','fontsize',12);
subplot(122)
imagesc(log10(SquareField_7_Frequency))
title(['Fourier Transform of a 2D Partial Checkerboard',...
    '(log of amplitude)'],'fontsize',12);
colormap(gray)

set(gcf,'Position',[40    300    900    400]);
```

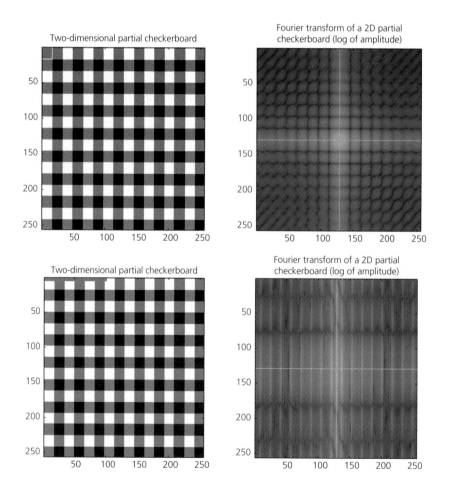

This introduction has set the ground to use the Fourier transform as a tool for analysing images of scratch wound assays.

5.9 Recipe 8: Filtering Scratch Wound Assays in the Fourier Domain

The Fourier transform has been widely used in image processing. Its popularity relies on the fact that some process like high or low pass filtering can be performed in the Fourier domain in the same way they can be done on the spatial domain. However, the real advantage comes from the fact that some processes on the spatial domain, like convolution or correlation, are more time consuming than their equivalent in the Fourier domain, which can be done by multiplication. Therefore, it is faster to transform from the spatial to the Fourier domain, filter in the Fourier domain and then transform back to the spatial domain, than to filter directly in the spatial domain. If one is to process a few small images, the advantage may not be relevant, but if there are numerous images to be processed, then we can consider the use of the Fourier transform. This recipe will perform some basic filtering in the Fourier domain, and the results will be similar to those obtained in previous recipes.

We can begin by reading an image of a scratch wound assay with different characteristics as the ones we have seen previously:

```
migration2 = imread('188t01.jpg');
```

The first step will be to obtain the two-dimensional Fourier transform with the command 'fft2' and then shift the positions so that the centre of the image corresponds to the location of 0 frequency. Since these images are grey scale and there is no relevant information in the colour, we can select the blue channel (or could be the red or green as well) to obtain the Fourier transform.

```
migrationImage_Frequency     = ((fftshift(fft2(migration2...
                                (:,:,1))))));
[rows,columns,levels]        = size(migrationImage_Frequency);
```

We have also calculated the size of the image, as it will be useful later. We can now display the scratch wound assay next to its Fourier transform. As we have learned previously, the very high intensity of the spikes can be difficult to observe as they can be much more intense than neighbouring elements, so we attenuate these by using a logarithm base 10.

```
figure
subplot(121)
imagesc(migration2(:,:,1))
title('Scratch Wound Assay','fontsize',14);

subplot(122)
imagesc(log10(abs(migrationImage_Frequency)))
title(['Fourier Transform of Scratch Wound Assay',...
```

```
    '(log of amplitude)'],'fontsize',14);
colorbar
colormap(gray)
set(gcf,'Position',[40    300    900    400]);
```

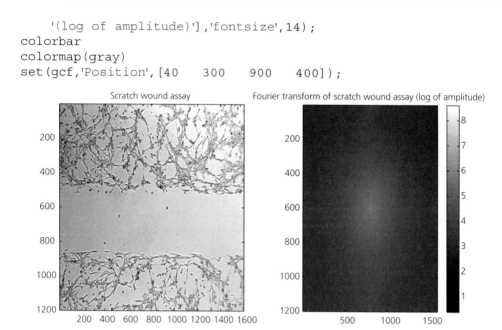

We can see now from the `'colorbar'` that the intensity spans from 0 to 8 and that the original Fourier transform of the image has spikes with magnitudes around 10^8. These spikes are concentrated around the central section of the image; this is the region of low frequencies. The region far from the centre is the region of the high frequencies. When we visualise the Fourier spectrum of an image, the terms "Low Pass" and "High Pass" become more intuitive as what we intend is to suppress a specific region of the spectrum to highlight the elements in the other regions.

To create a Low Pass filter we need to have a signal that will remove the spikes in the centre and leave the ones in the surrounding unaffected areas. A simple way of creating such a filter is to create a matrix of zeros of the same dimensions of the image, but with a few ones in the central region. This is called an "ideal filter" and, for several technical reasons, it is preferable to use other type of filters instead of ideal filters.

First we will filter the scratch wound assay with an ideal low pass filter. This will be zeros everywhere, except a region of 40 pixels to the left/right and up/down from the centre of the image. We emphasize that this and other filters are in the frequency domain on the name of the variable.

```
idealLowPassFilter_Frequency            = zeros(rows,columns);
idealLowPassFilter_Frequency...
    (rows/2-40:rows/2+40, columns/2-40:columns/2+40) = 1;
```

The filtering process is obtained by a pixel-by-pixel multiplication of the filter and the image in the frequency domain:

```
migrationImage_idealLPFiltered       = ...
    migrationImage_Frequency.*idealLowPassFilter_Frequency;
```

We can now display the filter, the filtered image in the Fourier and in the spatial domain next to each other.

```
figure
subplot(131)
imagesc(idealLowPassFilter_Frequency)
title('Low Pass Filter','fontsize',12);
subplot(132)
imagesc(log10(abs(migrationImage_idealLPFiltered)))
title('Filtered Frequencies','fontsize',12);

subplot(133)
imagesc(abs(ifft2(migrationImage_idealLPFiltered)))
title('Low Pass Filtered Image','fontsize',12);
colormap(gray)

set(gcf,'Position',[40    300    900    400]);
```

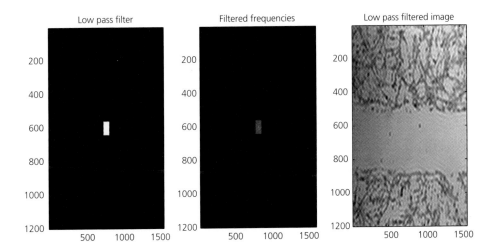

The ideal filter has removed, sharply, all the frequency components that are outside the small square region at the centre of the frequency spectrum. The corresponding filtered image in the spatial domain is a blurry version of the scratch wound assay. However, a close examination of the image reveals some artefacts. If we look at the region of the wound, which was pretty uniform in the original image, we can see that there are some wavy patterns.

```
figure
imagesc(abs(ifft2(migrationImage_idealLPFiltered)))
title('Low Pass Filtered Image','fontsize',12);
axis([ 200 500 400 700])
colormap(gray)
```

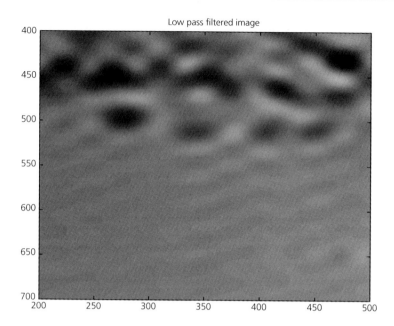

Low pass filtered image

These unwanted effects are consequence of the ideal filter. We will now apply a different filter, one which does not remove sharply the frequency components, but rather attenuates them gradually. These type of filters are preferable to the ideal ones and can be achieved through many mathematical functions, one of which is a Gaussian shape.

We can generate Gaussian shapes in several ways, for instance by programming the mathematical function itself, however, as in many other cases, Matlab provides a command to generate Gaussian as well as many other functions. This command is 'fspecial', and we need to specify the function, Gaussian, the dimensions, which will be the same as the image we are processing, and the width of the Gaussian function. A wider function will keep more low frequency components, a narrower function will attenuate more frequency components and the filtered results should be blurrier.

```
lowPassFilter_Frequency  = fspecial('Gaussian',...
                                     [rows columns],30);
```

It is important to notice that 'fspecial' will create a function that integrates to 1 in the two dimensions, that is, if we sum the values of all its elements along rows and columns:

```
sum(sum(lowPassFilter_Frequency))
```

```
ans =

1.0000
```

If we were using this filter in the spatial domain, we would need to keep the values of the filter to sum to one as the filtering process is performed by adding the values of neighbouring elements. However, in the frequency domain, we will multiply the filter by the image, and in this case we want the filter to have values between zero and one, so that the ones leave the frequency elements unchanged, and the zeros remove the frequency elements. Therefore, all we need to do is to divide the whole filter by its maximum value; this will scale all the intensity values to be between zero and one.

```
lowPassFilter_Frequency  = lowPassFilter_Frequency...
   /max(lowPassFilter_Frequency(:));
```

We can now filter the scratch wound assay figure in the frequency domain by multiplying it by the filter and then applying the inverse Fourier transform with the command 'ifft2'.

```
migrationImage_LPFiltered_Frequency = ...
   migrationImage_Frequency.*lowPassFilter_Frequency;
migrationImage_LPFiltered        = ...
   (ifft2(migrationImage_LPFiltered_Frequency));
```

It should be noted that Fourier transforms produce imaginary numbers, and we need to obtain their magnitudes with the command 'abs' to be able to display them properly.

```
figure
subplot(131)
imagesc(lowPassFilter_Frequency)
title('Low Pass Filter','fontsize',12);
subplot(132)
imagesc(log10(abs(migrationImage_LPFiltered_Frequency)))
title('Filtered Frequencies (log of ampl.)','fontsize',12);
subplot(133)
imagesc(abs(migrationImage_LPFiltered))
title('Low Pass Filtered Image','fontsize',12);
colormap(gray)

set(gcf,'Position',[40   300   900   400]);
```

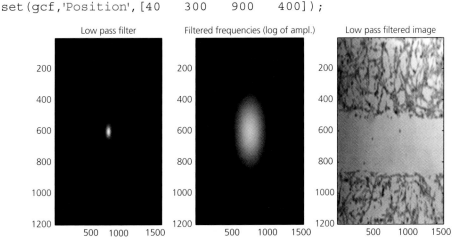

If we compare the two previous filters we can see that the size of the filters is not too different, however, the ideal low pass filter has completely removed any frequency component outside the central square. In comparison, the Gaussian filter attenuates gradually the frequencies from the centre, and thus there frequency elements that are not removed cover a much bigger area than previously. The general effect of the filter is again to smooth the image, but now we can see that the wavy artefacts that were present with the ideal filter have disappeared.

```
figure
imagesc(abs(migrationImage_LPFiltered))
title('Low Pass Filtered Image','fontsize',12);
axis([ 200 500 400 700])
colormap(gray)
```

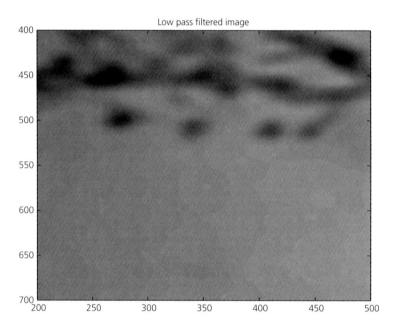

The low pass filter has removed the high frequency elements from the image and by doing so, it has produced a smooth, blurred version of the original image. The opposite effect, to remove the low frequency elements, can be achieved with a high pass filter, which we can easily obtain with the complement of the low pass filter previously created:

```
highPassFilter_Frequency    = 1-lowPassFilter_Frequency;
```

This filter will keep the high frequency elements and slowly attenuate the low frequency ones until the centre, where the value is zero (it was one in the previous filter). Thus, to filter the image, we only need to multiply the filter by the image in the Fourier domain:

```
migrationImage_HPFiltered    = migrationImage_Frequency.*...
    highPassFilter_Frequency;
```

As before we can display the filter, the filtered data and the inverse of the filtered data remembering that we need to use 'abs' to obtain the magnitude of the complex numbers, and we can use the logarithm as the individual values will be very high.

```
figure
subplot(131)
imagesc(highPassFilter_Frequency)
title('High Pass Filter','fontsize',12);
subplot(132)
imagesc(log10(abs(migrationImage_HPFiltered)))
title('Filtered Frequencies (log of ampl.)','fontsize',12);
colorbar
subplot(133)
imagesc(abs(ifft2(migrationImage_HPFiltered)))
title('High Pass Filtered Image','fontsize',12);
colormap(gray)
set(gcf,'Position',[40    300    900    400]);
```

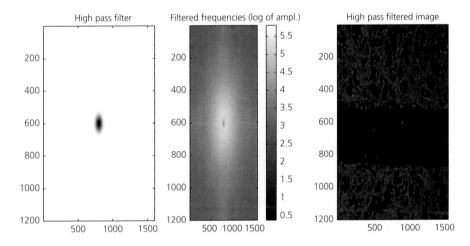

Notice that we have included a scale bar to compare the range of values against the values in the first image of this recipe, which were of the order of 10^8. In this case, since we have attenuated the low frequency elements of the image, which are the ones with higher intensities, the range is in the order of 10^5. This is a reduction of the magnitude by a factor of 1,000.

The filtered image has highlighted the regions where there are sharp transitions, i.e. edges where the intensity rises or drops considerably. On the other hand, the image has lost all the information related to the low frequency elements, that is the grey-level intensity of the background.

The low pass and high pass filters are just the entry point to a huge number of filters that can be obtained in the Fourier domain. For a start we can combine filters so that we remove low and high frequency elements at the same time, which is called a "band-pass filter" as only a band of frequencies is allowed through the filter. We will use the previous Gaussian filter and a second filter with a slightly different width to create the effect of removing the central frequencies as well as the outside ones.

```
lowPassFilter2_Frequency   = fspecial('Gaussian',...
                             [rows columns],25);
lowPassFilter2_Frequency   = lowPassFilter2_Frequency/...
                             max(lowPassFilter2_Frequency(:));

bandPassFilter_Frequency   = -lowPassFilter2_Frequency+...
                             lowPassFilter_Frequency;
migrationImage_BPFiltered = migrationImage_Frequency.*...
                             bandPassFilter_Frequency;


figure
subplot(131)
imagesc(bandPassFilter_Frequency)
title('Band Pass Filter','fontsize',12);
subplot(132)
imagesc(log10(abs(migrationImage_BPFiltered)))
title('Filtered Frequencies (log of ampl.)','fontsize',12);
subplot(133)
imagesc(abs(ifft2(migrationImage_BPFiltered)))
title('Band Pass Filtered Image','fontsize',12);
set(gcf,'Position',[40    300    900    400]);

colormap(gray)
```

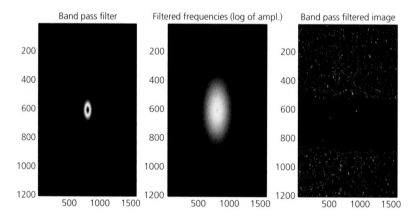

The filter becomes a ring around the centre, and the effect of this filter is similar to the high pass in the way that the image has no smooth background, but those regions that were highlighted by the filter are smoother (i.e. thicker lines) than those present in the high pass filter.

All the filters we have analysed so far have been symmetric around the centre, that is they attenuate, or not, the frequency components in the same way, regardless of the orientation from the centre. This does not need to be the case for every filter. We can filter only a section of the Fourier spectrum and obtain very different filtered images. This is the basis of a special type of filter called "Gabor" filters that are used to subdivide in frequencies and orientation the spectrum, and are widely used to analyse texture from images.

Gabor filters use Gaussian functions that are not placed at the centre of the spectrum. For simplicity, we will illustrate the orientation of a filter with ideal filters. We will create first a filter that divides the spectrum into quadrants, upper left (U.L.), upper right (U.R.), bottom left (B.L.) and bottom right (B.R.) and removes the frequency components in the U.R and B.L. quadrants.

To create this filter we will need a matrix of zeros of the same size as the original image, and we will place ones at the corresponding quadrants.

```
quadrantFilter_1   = zeros(size(migrationImage_Frequency));
quadrantFilter_1(1:rows/2,1:columns/2)           = 1;
quadrantFilter_1(rows/2+1:end,columns/2+1:end)   = 1;
```

We can now filter the image and display the results:

```
migrationImage_ULBRQuadrant = ...
    migrationImage_Frequency.*quadrantFilter_1;
```

```
figure
subplot(131)
imagesc(quadrantFilter_1)
title('Quadrant Filter (U.L.,B.R.)','fontsize',12);
subplot(132)
imagesc(log10(abs(migrationImage_ULBRQuadrant)))
title('Filtered Frequencies','fontsize',12);
subplot(133)
imagesc(abs(ifft2(migrationImage_ULBRQuadrant)))
title('Quadrant Filtered Image (U.L.,B.R.)','fontsize',12);
colormap(gray)
set(gcf,'Position', [40    300    900    400]);
```

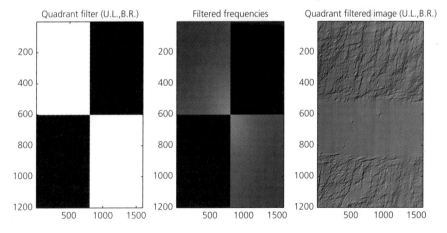

The effect of this quadrant filter is very different from all the other filters. For a start the background of the image is now an intermediate level of grey, and the variation due to the cells is indicated as lower or higher intensities, similar to the effects of interference contrast images. This type of filters are sometimes called "embossing" filters. Since we have used ideal filters we can notice that there are some undesired artefacts, which are visible in the central part of the image.

```
figure
imagesc(abs(ifft2(migrationImage_ULBRQuadrant)))
colormap(gray)
axis([350 650 480 880])
```

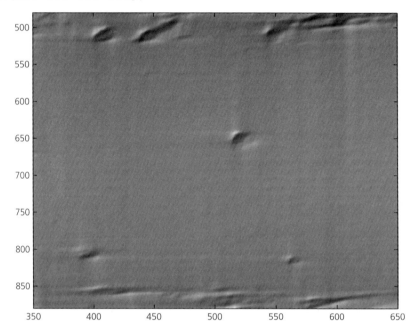

The artefacts are manifest as horizontal and vertical "shadows" that span from the cells on the image.

The embossing effect of this filter arises from the fact that the filter is including a very special element, the one at the "centre", or more precisely, since the image has an even number of rows and columns, at the right-bottom side of the centre. This point is called the "zero frequency" element, or in some cases the "D.C. component" in reference to the direct current as opposed to alternate current. If we remove that single element by shifting the filter by one column we will get a very different filter:

```
quadrantFilter_2   = zeros(size(migrationImage_Frequency));
quadrantFilter_2(1:rows/2,1:columns/2)             = 1;
quadrantFilter_2(rows/2+1:end,columns/2+2:end)   = 1;



migrationImage_ULBRQuadrant_2 = ...
    migrationImage_Frequency.*quadrantFilter_2;

figure
subplot(131)
imagesc(quadrantFilter_2)
```

```
title('Quadrant Filter (U.L.,B.R.)','fontsize',12);
subplot(132)
imagesc(log10(abs(migrationImage_ULBRQuadrant_2)))
title('Filtered Frequencies','fontsize',12);
subplot(133)
imagesc(abs(ifft2(migrationImage_ULBRQuadrant_2)))
title('Quadrant Filtered Image (U.L.,B.R.)','fontsize',12);
colormap(gray)
set(gcf,'Position',[40    300    900    400]);
```

We can now see that the grey background has disappeared as this was contained in that zero-frequency element. The artefacts are still present as these are related to the high frequency elements of the image.

We can also obtain the complementary filter, one that removes the frequencies in the upper-left and bottom-right quadrants:

```
quadrantFilter_3    = zeros(size(migrationImage_Frequency));
quadrantFilter_3(1:rows/2,columns/2+1:columns)    = 1;
quadrantFilter_3(rows/2+1:end,1:columns/2)         = 1;

migrationImage_URBLQuadrant =...
    migrationImage_Frequency.*quadrantFilter_3;
figure
subplot(131)
imagesc(quadrantFilter_3)
title('Quadrant Filter (U.R.,B.L.)','fontsize',12);
subplot(132)
imagesc(log10(abs(migrationImage_URBLQuadrant)))
title('Filtered Frequencies','fontsize',12);
subplot(133)
imagesc(abs(ifft2(migrationImage_URBLQuadrant)))
title('Quadrant Filtered Image (U.R.,B.L.)','fontsize',12);
colormap(gray)
set(gcf,'Position',[40    300    900    400]);
```

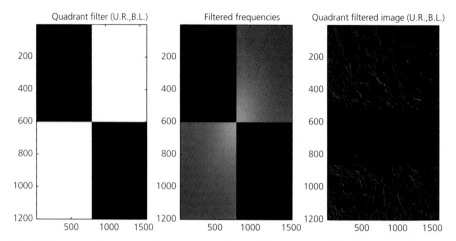

The differences between the two quadrant filters are subtle but very important. The following figure zooms in to a special region of interest of the image to show these differences:

```
figure
subplot(131)
imagesc(migration2(:,:,1))
axis([300 680 820 1200])
title('Original Image','fontsize',12);

subplot(132)
imagesc(abs(ifft2(migrationImage_ULBRQuadrant_2)))
axis([300 680 820 1200])
title('Quadrant Filter (U.L.,B.R.)','fontsize',12);

subplot(133)
imagesc(abs(ifft2(migrationImage_URBLQuadrant)))
axis([300 680 820 1200])
title('Quadrant Filtered Image (U.R.,B.L.)','fontsize',12);
colormap(gray)
set(gcf,'Position',[40    300    900    400]);
```

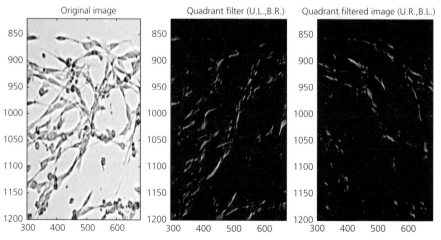

The two filtered images now present a complementary behaviour in highlighting cells with different orientations. The central image is highlighting those in a diagonal from the bottom left to the top right whilst the image on the right has highlighted the cells on the diagonal from the bottom right to the top left.

These are very rough filters, but it is not too difficult to partition the frequency spectrum in search of regions of specific frequency and orientation with the objective of segmenting regions of different textures.

Further description of these filters, sometimes called sub-band filters, is beyond the scope of this chapter. The publications by Randen and Husoy and Reyes-Aldasoro and Bhalerao contain more details.

References

Otsu, Nobuyuki. 1979. "A Threshold Selection Method from Gray-Level Histograms." IEEE Transactions on Systems, Man and Cybernetics 9, No. 1 (January): 62–66. doi:10.1109/TSMC.1979.4310076.

Randen, Trygve, Husoy, John Hakon. 1999. "Filtering for texture classification: a comparative study." IEEE Transactions on Pattern Analysis and Machine Intelligence 21 (4), 291–310. doi: 10.1109/34.761261.

Reyes-Aldasoro, Constantino Carlos, Bhalerao, Abhir. 2006. "The Bhattacharyya space for feature selection and its application to texture segmentation." Pattern Recognition 39 (5), 812–826. doi: 10.1016/j.patcog.2005.12.003.

CHAPTER 6

Bright Field Microscopy

Bright Field Microscopy is the simplest and one of the most common techniques used in light microscopy and is widely used to observe cells "in vitro", slices of tissue or live organisms. The samples of interest are placed on the stage of a microscope, and they are illuminated with white light that passes through the sample and then the light travels through a series of lenses that form the microscope until the light beams reach the eye of the observer or a camera where an image is recorded. In some cases, it is possible to illuminate the sample from the same side of the microscope lenses and therefore the light collected is the reflection from the sample. Other microscopic techniques involve using fluorescent substances or illuminating the samples sideways to exploit the contrast of the image. We will not cover those in this section.

This imaging technique is a tried-and-tested method, relatively simple and widely used. The applications of image analysis with images acquired through bright field are many, from the correction of shading to the stitching of images to form a larger image to segmentation and measurement of cells or other objects of interest. The recipes in this section will cover several of these applications.

6.1 Recipe 1: Changing the Brightness and Contrast of an Image

Probably one of the most common tasks of image analysis is that of adjusting brightness and contrast of images. It may be the case that images were acquired in different days and the settings of the microscope were different so that one group of images is darker than another. The images may come from different equipment or simply some images are too bright or dark and we would like to change them.

In this recipe we will explore different ways in which the brightness and contrast of an image can be manipulated. We will aim to create images with better visual appearance, and also, with a larger range of intensities so that later processing steps can benefit from an enhanced image.

First, we need a sample image. For this example we will read an image of some cells stained through immunohistochemistry. We use the command 'imread' to

Biomedical Image Analysis Recipes in MATLAB®: For Life Scientists and Engineers, First Edition.
Constantino Carlos Reyes-Aldasoro.
© 2015 John Wiley & Sons, Ltd. Published 2015 by John Wiley & Sons, Ltd.

read the image, which in this case is located on a web site, so we pass the complete web site address in between single quotes as argument to 'imread'. The output of 'imread' is saved in the variable 'sampleImage'. We will also calculate the size of the image for further manipulations with the command 'size' and save those values in three separate variables.

```
sampleImage = imread(...
  'http://caiman.group.shef.ac.uk/caiman/images/shading2.jpg');
[rows, columns, channels]= size(sampleImage);
```

Once we have the image in Matlab, we can display it with the command 'imagesc':

```
figure
imagesc(sampleImage)
```

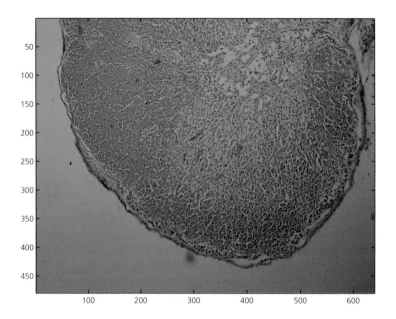

Image courtesy of Professor Gillian Tozer, The University of Sheffield, UK.

Notice that we have created a new figure so that the image is displayed in this figure. It is possible to use a single figure and overwrite the images to be displayed, but in some cases it is convenient to keep previous results to be able to compare the effects of the steps of an algorithm.

The intensity of the image is given by the values of its corresponding pixels; therefore this intensity can be easily manipulated by increasing or decreasing the values of the individual pixels. For example, we can add or subtract a constant value to all pixels of the image or to just a few. To illustrate this we will create a copy of the sample image and add an arbitrary value to the top of the image and we will subtract that value to the bottom of the image.

```
sampleImage2                          = sampleImage;
sampleImage2(1:rows/2,  :,  :)        = sampleImage2...
   (1:rows/2,:,:)+30;
sampleImage2(rows/2+1:end,  :,  :) = sampleImage2...
   (rows/2+1:end,:,:)-30;

figure
imagesc(sampleImage2)
```

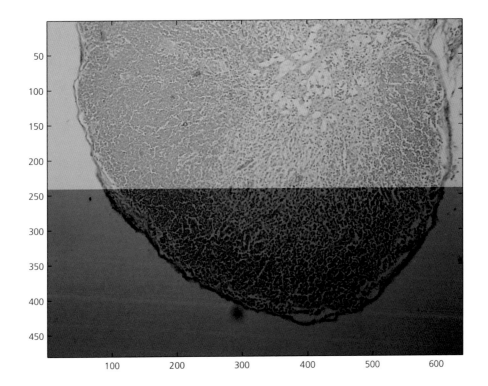

Alternatively, if we do not wish to create a new variable, we can concatenate two or more matrices by placing them inside square brackets ([ ]). Therefore, we can concatenate half of the image, say the left side, to which we add a constant, and the right side to which we subtract the constant.

```
figure
imagesc([sampleImage(:,1:columns/2,:)+80 ...
        sampleImage(:,columns/2+1:end,:)-80])
```

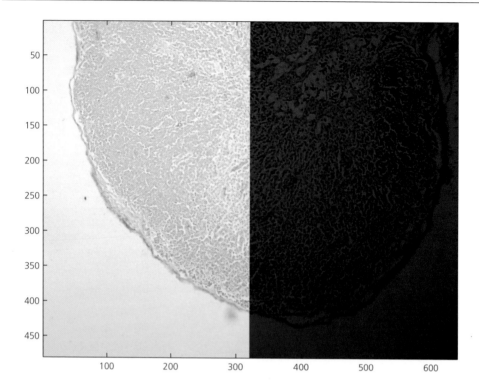

By increasing or decreasing the intensity values, we can make an image brighter or darker as can be seen in the previous images. The values of the pixels of the images that we have displayed so far are of the 'uint8' ("un-signed integer of 8 bits") data type, and therefore the minimum and maximum possible values are 0 and 255, respectively. Say the original value of a pixel is 200 and we add 70, the output would be capped at 255; equally any negative values are set to 0. This limiting process is called "saturation" of the values.

Notice how we have addressed the matrices: in the first case we selected a subset of the rows, all the columns and all the channels with the use of the colon operator (":"), and in the second case all rows, channels and a subset of the columns.

Notice also that when a matrix is addressed, the values of the coordinates should always be positive and integers. In this case, the number of columns is an even number, and thus when divided by 2 the division is an integer as well, but if the number of columns would be an odd number, then we must avoid the decimals by using commands like 'floor', 'ceil' or 'round' that round a number down, up, or to the nearest integer, respectively. Look at these examples:

```
[floor(4.1) ceil(4.1) round(4.1);...
   floor(4.9) ceil(4.9) round(4.9);]

   ans =

      4      5      4
      4      5      5
```

Now we will investigate how to increase the contrast of the image, which has the effect of making the dark regions darker, and the bright regions brighter.

The increase in contrast can be obtained in different ways with mathematical operations. However, Matlab provides a command that can automatically increase contrast. The command is 'imadjust', which stands for "image adjust". For simplicity we will explore the command for grey-scale images before we explore the use of this command for colour images. Thus, we will convert the colour image from RGB to grey scale with the command 'rgb2gray', and then adjust the contrast like this:

```
sampleImage_grey              = rgb2gray( sampleImage);
sampleImage_grey_adjusted     = imadjust(sampleImage_grey);
```

The first input argument for the command 'imadjust' is the image whose contrast is to be adjusted. We can use other arguments, which will be presented below.

We can now compare the original image and the adjusted one, again by showing the original next to the adjusted together.

```
figure
imagesc([sampleImage_grey(:, 1:columns/2)...
    sampleImage_grey_adjusted(:, columns/2+1:end)])
colormap(gray)
```

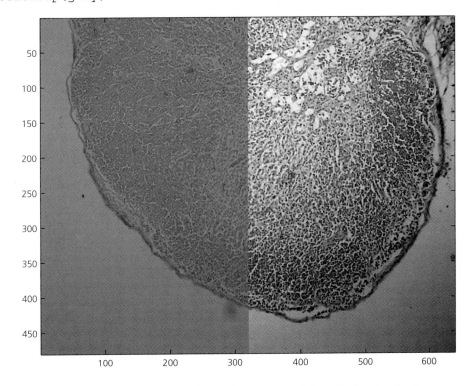

In addition we can display the intensity values of the pixels of a single row next to each other.

```
figure
xRow=100;
plot((1:columns/2), sampleImage_grey(xRow, 1:columns/2),...
    'r-',(columns/2+1:columns),...
    sampleImage_grey_adjusted(xRow,columns/2+1:end));
grid on;
axis([1 columns 0 255])
```

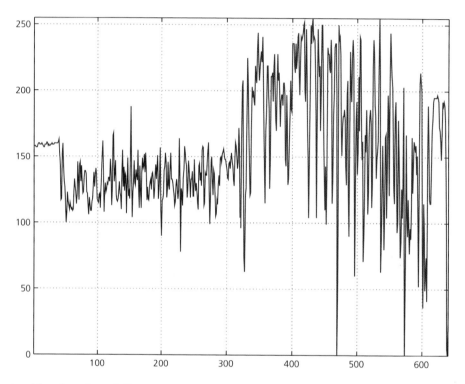

Besides the clear difference in contrast between the two sides of the image, the difference of the intensities in the row is striking; it is clear that the intensities now spread from 0 to 255.

It is possible to use the command 'imadjust' with colour images. However, it is useful to know a bit more about the nature of the images, as some parameters are required to adjust the intensity levels of the channels of the image. We can investigate the intensities of the image with the aid of a histogram displayed with the command 'hist'. Since we are interested in the distribution of the pixels over the horizontal axis, we will add a grid in the figure with divisions or "ticks" spaced every 20 grey levels.

```
figure
hist(double(sampleImage(:)), 100)
xlabel('Pixel Intensity')
ylabel('Pixel Occurrence')
```

```
set(gca, 'xtick', (0:20:250))
grid on
axis tight
```

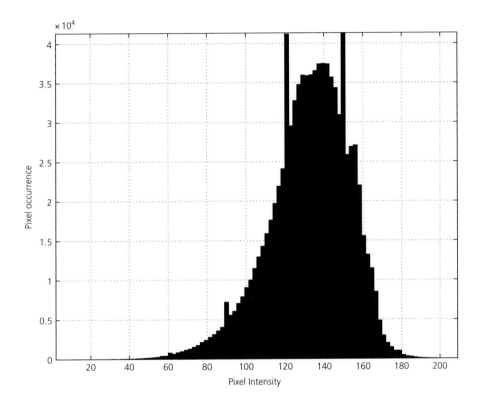

Notice that it is necessary to convert the values from the 'unit8' data type to 'double' to calculate the histogram. The histogram shows that the majority of pixels of the image are located within the range of 60 to 180, with very few pixels in the extremes outside these values. In this particular example, we can see that the highest intensity of the image is around 220. It would make sense then to stretch the intensity values so that the highest value is shifted to 255, which is the maximum value of a 'uint8' data type. However, we could also argue that since there are very few pixels in the region 180–220, we could shift the values so that 180 becomes 255. All the values above 180 would then be capped at 255; these pixels would be "saturated". The equivalent transformation on the dark side of the histogram would be to shift from a given value, say 40 to 0.

Based on the histogram, we can select a low and a high value (say 60 and 180 for the present example), which will become the new extremes, with black represented by 0 and white represented by 255. These values are passed as a second (60,180) and third arguments (0,255) to 'imadjust'. Since 'imadjust' can be used for different data types, the low and high values have always normalised to the range [0–1] so we divide the values 60 and 180 by 255 in the following way:

```
sampleImage_adjusted = ...
   imadjust(sampleImage,[60,180]/255,[0,255]/255);
```

Now we can observe the adjusting of the contrast, again showing half of the original image and half of the adjusted one:

```
figure
imagesc([sampleImage(:, 1:columns/2, :) ...
    sampleImage_adjusted(:, columns/2+1:end, :)])
colormap(gray)
```

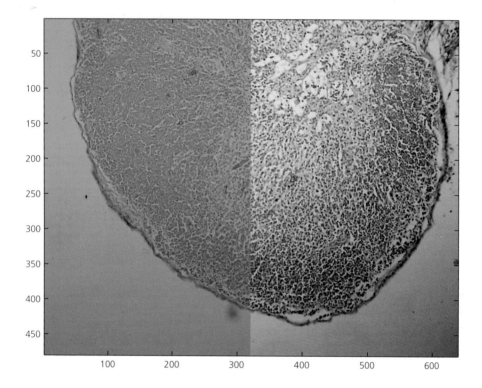

We can increase (or decrease) the contrast by raising (or lowering) the lower value and lowering (or raising) the higher value, which results in an image with higher (or lower) saturation. For example, we can select (100,140) as the new low and high values and observe the new adjusted image.

```
sampleImage_adjusted = ...
   imadjust(sampleImage,[100;140]/255,[0;1]);

figure
imagesc([sampleImage(:, 1:columns/2, :)...
    sampleImage_adjusted(:, columns/2+1:end, :)])
colormap(gray)
```

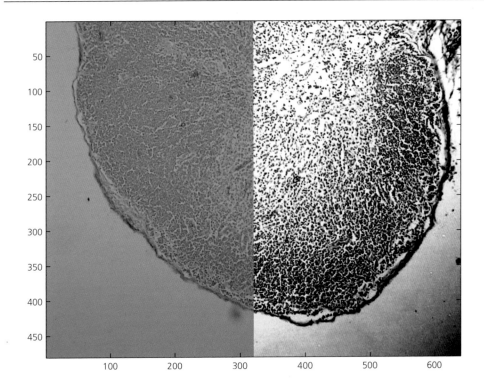

The previous adjustments were applied uniformly to the three colour channels. However, it is possible to apply different levels of adjustment to each channel by passing different low and high levels. In this way, what we are deciding is how to stretch each channel, by selecting a low and high value for each channel.

The following example shows changing the contrast for the three channels separately:

```
sampleImage_adjusted2  = imadjust(sampleImage, ...
                         [90 60 60; 140 180 180]/...
                         255,[0 0 0; 1 1 1]);
sampleImage_adjusted3  = imadjust(sampleImage, ...
                         [60 90 60; 180 140 180]/...
                         255,[0 0 0; 1 1 1]);
sampleImage_adjusted4  = imadjust(sampleImage,...
                         [60 60 90; 180 180 140]/...
                         255,[0 0 0; 1 1 1]);

figure
imagesc([sampleImage_adjusted sampleImage_adjusted2;...
    sampleImage_adjusted3 sampleImage_adjusted4])
```

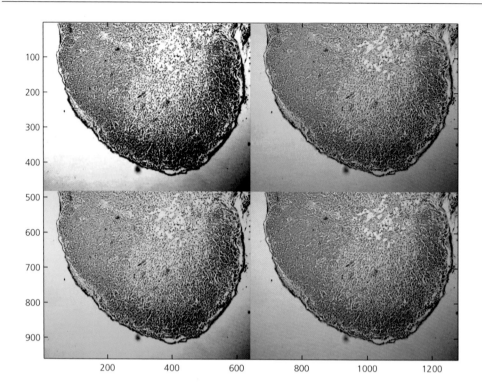

It is interesting to notice how stretching channels separately can not only increase contrast, but shift the colour of the image, especially of the background.

A final observation of the image is related with the next recipes. We can see that the intensity of the image, especially the background where there are no cells, is not uniform. This phenomenon is known as intensity inhomogeneity or shading, and it is aggravated by the increase of the contrast. The following recipes explore methods to deal with shaded images.

6.2   Recipe 2: Shading Correction: Estimation of Shading Component as a Plane

The process through which the image is formed, and its resulting quality, depend on the conditions of several of the stages of the path of the light. If the sample, for instance a slice of tissue, is not uniformly dense, or the thickness varies, the light will experiment different attenuation levels at different locations of the image. Bulbs also present a challenge because they do not always emit light uniformly. This can be partly compensated by the optimal arrangement of the optics of a microscope, commonly known as Koehler illumination. It is common to observe images in which the intensity is not uniform. This may not be important for images that will be visually examined. However, when a series of images are to be processed automatically, there are important implications if the intensity is not uniform. For example, we can analyse the following slide of a tumour stained by immunohistochemistry:

```
shadedImage = imread(...
  'http://caiman.group.shef.ac.uk/caiman/images/shading2.jpg');
figure
imagesc(shadedImage)
```

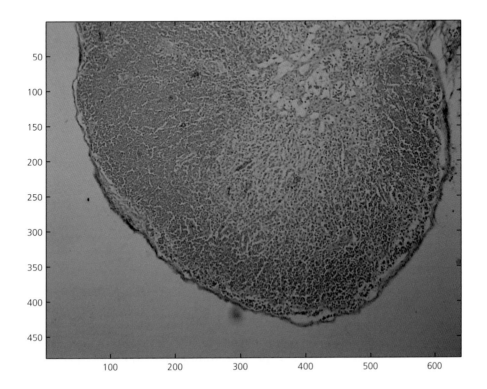

The staining of the image is fairly uniform, however, the nonuniform intensity is easily recognizable in the areas of the image where there are no cells. The bottom corners have a darker shade than the upper regions without cells. To explore the intensity levels a step further, we can plot a single row of the intensity. Since the image is a colour image with three channels for red, green and blue, we can plot the intensity of each channel on the same plot. This is an important step, as in some cases, the optical settings may be introducing different bias to each channel.

First, we need to obtain the size of the image, as we will use the dimensions of the rows and the columns in subsequent steps.

```
[rows, columns, channels]         = size(shadedImage);
```

Since this is a colour image, we know that it will have three channels. There are some cases in which colour images may have four channels as they may contain transparency and other characteristics. Next, we select two rows of the image to plot, one from the middle of the image and one from the bottom. We can arbitrarily select rows 300 and 400.

To visualise the rows of the image that we have selected, we can overlay a black line over those rows. To do this, we can copy the original image to a new one, and set to 0 the pixel values of those rows. In general it is a good idea to keep the original data untouched so that it can always be returned to if necessary. To make the lines thicker in the plot we have set to 0 four rows in each case.

```
shadedImage2                = shadedImage;
shadedImage2(300:303, :, :) = 0;
shadedImage2(400:403, :, :) = 0;
```

To display the modified image we use the command 'imagesc'.

```
figure
imagesc(shadedImage2)
```

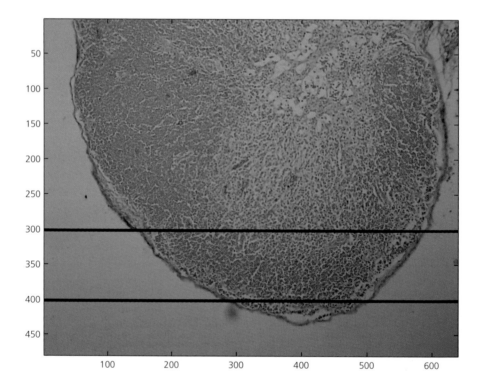

To display the intensity of the pixels over those rows, we use the command 'plot', and we will pass as arguments the following: the horizontal axis formed by a 1D matrix (1:columns), the vertical axis formed by the actual values of the pixels and the characteristics of the line. Since we want the three channels in the same plot, we will select one channel per line by addressing the third dimension of the matrix, e.g. 'shadedImage(xRow, :, 1)'. In this way, we have selected the values of the row 'xRow', all the columns (:) and the first channel.

```
xRow=300;
figure
plot((1:columns), shadedImage(xRow, :, 1), 'r-', ...
     (1:columns), shadedImage(xRow, :, 2), 'g--', ...
     (1:columns), shadedImage(xRow, :, 3), 'b-.', ...
     'linewidth', 2);
grid on;
axis([1 columns 0 255])

xRow=400;
figure
plot((1:columns), shadedImage(xRow, :, 1), 'r-', ...
     (1:columns), shadedImage(xRow, :, 2), 'g--', ...
     (1:columns), shadedImage(xRow, :, 3), 'b-.', ...
     'linewidth', 2);
grid on;
axis([1 columns 0 255])
```

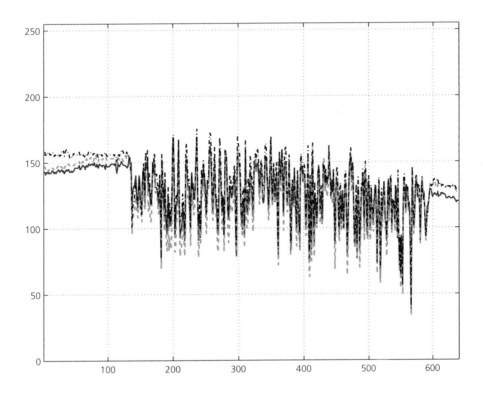

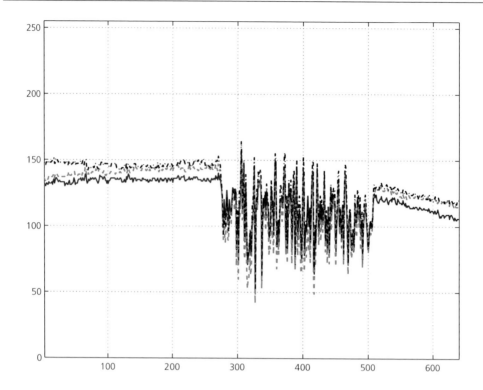

These plots are commonly referred to as profile plots, as they are similar to topographical profiles in which the height of a region is displayed.

In both rows it can be seen how the intensity of all channels decreases towards the right-hand side of the plot. It is also important to notice that not all channels behave in the same way: in the left-hand columns, the red and green channels tend to increase the intensity slightly whilst the blue channel is decreasing. Overall these differences between channels are rather small compared with the overall shading. In the next recipes we will introduce different techniques to deal with the shading.

When an image has shading we will try to correct this shading by creating an image with uniform pixel intensity. To correct the shading, it is necessary to make some assumptions, i.e. we need to speculate the reason for which the intensity is not homogenous along the image. One common assumption is to consider that a perfect, uniform image (we can call it "U") was obscured by a shading component ("S"), which may be due to the light, the optics or the sample itself. We may further assume that the shading has obscured the image in an additive way so that the actual image ("I") can be expressed as: $I = U + S$. Therefore if we can calculate "S", an approximation of U can be obtained from I-S. How good the approximation is will depend on how good the calculation of "S" is.

The process of estimating the shading "S" will require us to constrain "S" in a way in which we can modify it mathematically, i.e. a mathematical model of the shade. A good starting point for modelling problems is to visualise whatever we want to model. In our case, we can have a good view of the shade if we display the image as a 3D mesh. To do this, we need to either show one channel or a combination of the channels. For simplicity, we will assume that there is only one shading component, common to all three channels. Thus, we convert the colour RGB (Red, Green, Blue) image to grey scale with the command 'rgb2gray' and at the same time will convert the values from the 'uint8' data type to 'double'.

```
shadedImage_grey          = double(rgb2gray( shadedImage));
```

Next, we display the image as a mesh as follows:

```
figure
mesh(shadedImage_grey)
view(-140, 25)
colorbar
axis tight
```

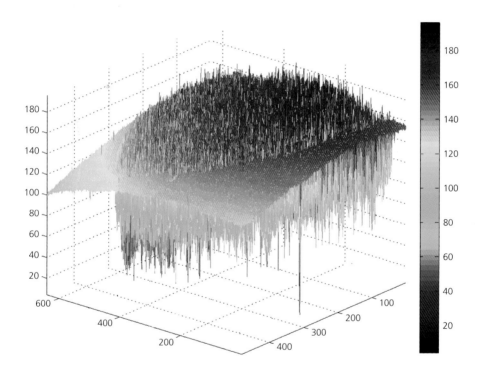

From the figure it is clear how the intensity is increasing towards the right-hand side of the plot.

The next step is to estimate the shading component. The estimation of a surface can be a complicated mathematical process, especially when a good estimation is expected. Techniques to fit curves or warp surfaces exist, and Matlab has toolboxes to do this, but this is beyond the scope of this book. For simplicity, we will assume that the shading component is a simple plane, e.g. a flat surface.

One way to estimate the plane that fits the shading component closely is to calculate the plane as the sum of two planes: one plane that increases only in one dimension, say the columns whilst the intensity over the rows is constant, and another planes that increases on the other dimension.

To calculate one plane, we can start by calculating one line; this line will be determined by the intensities of the extreme points of one dimension of the surface. Then we will replicate this line many times to form a plane. The other plane will be obtained in the same way but from the opposite dimension.

To generate the horizontal line we can select the intensities of the top corners of the image, that is, the first and last columns of the first row. Then we generate the line by creating a linearly spaced vector with the command 'linspace' like this:

```
horizontalLine = linspace(shadedImage_grey(1, 1),...
                shadedImage_grey(1, end), columns);
```

There are several ways to convert the horizontal line into a plane. One of them is to repeat the horizontal line, which was obtained from one row, to be repeated so that the plane has the same number of rows as the original image. We can do this with the command 'repmat' (repeat matrix) like this:

```
horizontalPlane          = repmat(horizontalLine, rows, 1);
```

Alternatively, we can perform a matrix multiplication between the horizontal line and a matrix of ones:

```
horizontalPlane          = ones(rows, 1)*horizontalLine;
```

Both instructions produce the same result. We can now obtain the complementary plane in the vertical dimension in the same way. This time we use the extreme values of the last column of the image to create the line, and then we use the matrix multiplication to generate the plane:

```
verticalLine  = linspace(shadedImage_grey(1, end),...
                shadedImage_grey(end, end), rows);
verticalPlane = verticalLine'*ones(1, columns);
```

We can visualise the two planes in the following figure.

```
figure
subplot(121)
mesh(horizontalPlane)
view(-140, 25)
axis tight

subplot(122)
mesh(verticalPlane)
view(-140, 25)
axis tight
```

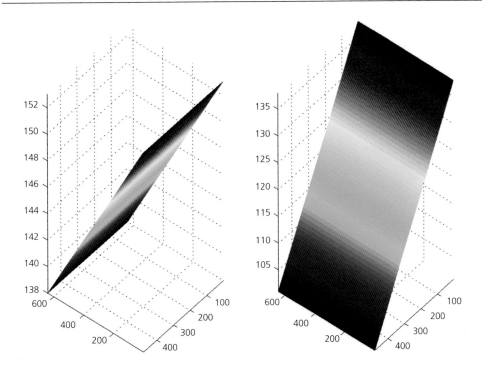

The two planes correspond to the variations of one single horizontal row and one single vertical column. It must be noted that this worked well for this image as the corners that were selected were part of what is supposed to be a uniform background. It may not be as simple in other cases. Nevertheless, this is the simplest of the approaches for shading correction; we will explore more complex methods later on.

Once we have the two planes, we can combine them as a simple average of the two planes to have our estimated shading component.

```
combinedPlane                = 0.5*horizontalPlane+0.5* ...
                               verticalPlane;
```

As was mentioned previously, before we proceed to correct the shading of the image, it is useful to visualise the intermediate steps. In that way we can visually evaluate how well this plane is approximating the shading that we perceive from the image. To do this, we plot together the intensity of the image and the plane:

```
figure
mesh(combinedPlane(1:4:end, 1:4:end, 1), 'edgecolor', 'b')
hold on
mesh(shadedImage_grey(1:4:end, 1:4:end, 1))
view(-140, 25)
axis tight
```

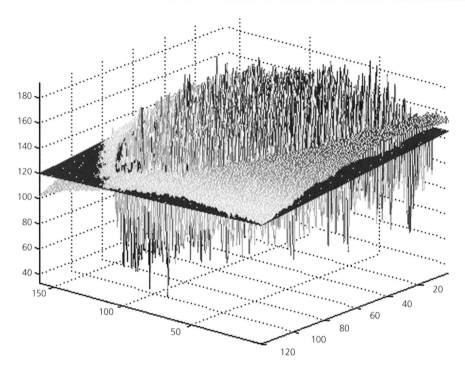

Notice that we did not plot all the points in the plane and the shading, we subsampled by taking one out of every four points. Subsampling is a common technique used with larger data sets; it saves time and is less intensive for the computer. Notice that the dimensions on the axes are also reduced.

It can be seen that the plane is roughly on the same intensity region as the image. On the left-hand side corner, the plane has higher intensities than the image. Towards the right-hand side, the plane has lower intensities. Thus, the estimation is not perfect, but it is close to the intensities we are trying to correct.

To correct the image from its shading, we need two more steps. First, the plane has the same range of intensities as the data. For the correction we want some intensities of the image to increase, i.e. the dark regions, some to remain constant and some to decrease, i.e. bright regions. For this to happen, the shading estimation needs to be, not in the same range of intensities as the data, but centred at 0 and have some positive and negative regions. We then need to shift this plane down by subtracting the mean value of the plane. The second step is to repeat the plane three times, one for each colour channel. This is done again using the command 'repmat' to repeat the matrix three times. We can do these two steps in one single instruction line in Matlab.

```
shadingComponent = ...
    repmat(combinedPlane-mean(combinedPlane(:)),[1 1 3]);
```

Finally, we can subtract the estimated shading component from the original image to generate a corrected image and we can display the outcome.

```
correctedImage_plane = ...
    uint8(double(shadedImage)-(shadingComponent));
```

```
figure
imagesc(correctedImage_plane)
```

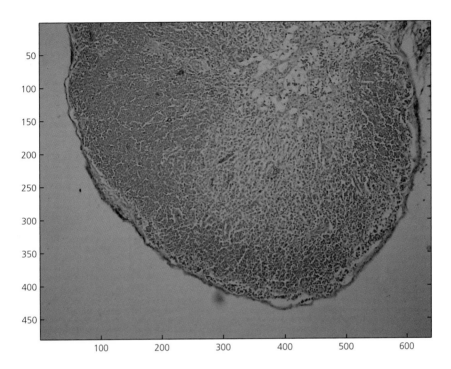

Notice that we included the word "plane" in the name of the variable, which is to indicate the method that we used to correct the shading.

From a visual assessment it is difficult to evaluate from the image how good the correction is. We can therefore display two profiles for the same rows as we had done before.

```
figure
xRow=300;
plot((1:columns), correctedImage_plane(xRow, :, 1), 'r-', ...
    (1:columns), correctedImage_plane(xRow, :, 2), 'g--', ...
    (1:columns), correctedImage_plane(xRow, :, 3), 'b-.', ...
    'linewidth', 2);
grid on;
axis([1 columns 0 255])
xRow=400;
figure
plot((1:columns), correctedImage_plane(xRow, :, 1), 'r-',...
    (1:columns), correctedImage_plane(xRow, :, 2), 'g--',...
    (1:columns), correctedImage_plane(xRow, :, 3), 'b-.',...
    'linewidth', 2);
grid on;
axis([1 columns 0 255])
```

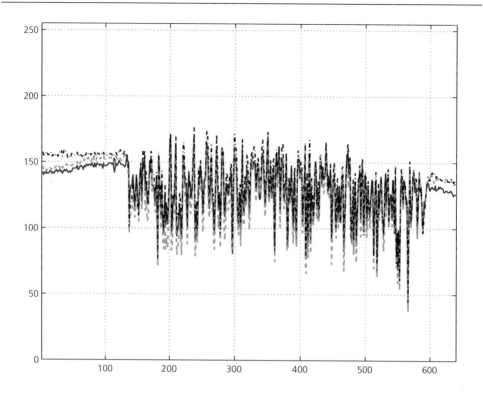

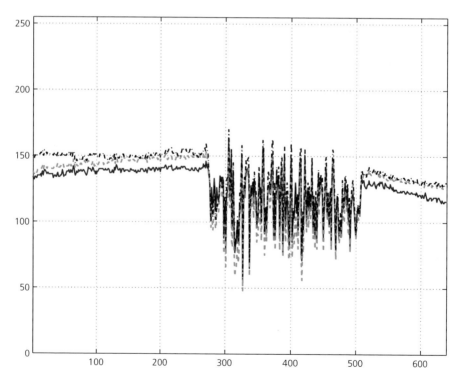

The profiles show how for both rows, the intensity of the left-hand side is now similar to the right-hand side. It is also clear that the intensity is still presenting some element of shading, which is revealed as the intensity increases towards the centre of the plot for the intensities of row 100, and then decreases towards the right-hand side. This indicates that a plane is not the best way to approximate the shading component of this image; this is not surprising, as we had observed that the shading component of the image was curved.

Also, it can be seen that the intensity of the blue channel is higher on the left side of the plot than the other two channels. Since this is a background region, it should clearly be a white (or rather grey) shade, and the intensities of all the channels should be equal. This indicates that a single shading component for all channels is not the best way to correct the shading.

6.3   Recipe 3: Estimation of Shading Component with Filters, Morphological Operators and Envelopes

In the previous recipe, we tried to remove the shading component of an image by assuming that the shading was approximately a flat surface. Whilst this technique partially improved the shading, there are other ways of estimating the shading component. We will examine some other methodologies in this recipe.

The first technique to estimate the shading component relies on the fact that the pixels themselves have been affected by the shading, and therefore the intensities are not only due to the nature of data itself, i.e. the cells of the stained tissue, but also the shading that corrupted the image. Thus, if we combine the intensities of neighbouring pixels, we can minimise the contribution of the data to the image and extract the effect of the shading itself. To combine the intensities of neighbouring pixels, we apply a common technique called "filtering", in which the intensities of the pixels of the image are combined according to another matrix, sometimes called a filter or a kernel. In this case, we will use the simplest filter, one with equal values in a square shape of size 15×15.

```
filterImage            = ones(15)/(15*15);
```

The values of the filter have been normalised so that the sum of all the elements of the filter adds to one. In other words, we are obtaining an average value of neighbouring pixels and replacing the value of one pixel by that of the average of the intensity of its neighbours. To filter the image we use the command 'imfilter' (image filter) and we pass as arguments, the image to be filtered, the filter which will be used, in this case the matrix 'filterImage', in the following way:

```
shadingComponent_filter = imfilter(shadedImage_grey,...
   filterImage,'replicate');
```

The third argument 'replicate' indicates to 'imfilter' how to deal with the edges of the image. This is important, as there are various options: not filter the edges, pad the edges with zeros, or pad the edges with the same value of the edge, i.e. to replicate the edge. By "padding" we mean that the image is going to grow in dimensions in all directions, and new pixels will be "padded" there. Thus, after the padding it is possible to apply the filters to the pixels of the edge of the original image as they will be somewhere on the inside of the padded image. In this example we prefer to replicate the value of the edge, but it is a good experiment to change this option and observe the differences.

Since this filter is giving equal weight to all neighbours, the effect is a general smoothing of the image. This is also known as a Low Pass Filter.

```
figure
mesh(shadingComponent_filter(1:4:end, 1:4:end, 1))
view(-140, 25)
axis tight
```

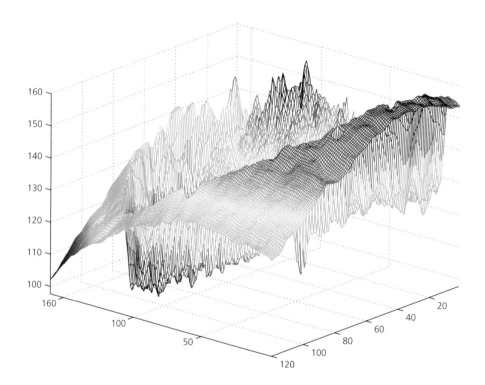

The mesh now shows the estimation of the shading component. It can be seen that for the regions of background, the mesh follows very closely the intensity, and therefore the estimation of the shade is very good. Where the image contains cells, the shade increases or decreases according to the intensity of those regions, therefore the approximation will not be that good in those regions.

We can compare this estimation with the plane by plotting them together:

```
hold
mesh(combinedPlane(1:4:end, 1:4:end, 1), 'edgecolor', 'b')
```

> *Current plot held*

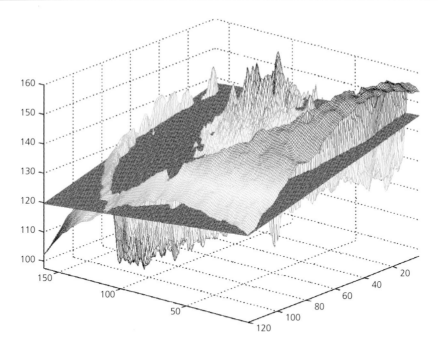

To finish the correction we estimate the inverse of the shading component, replicate it to the three channels, subtract it from the original image and display the image and the profiles as before:

```
inverseShade_filter    = repmat(shadingComponent_filter-...
                         mean(shadingComponent_filter(:)),...
                         [1 1 3]);
correctedImage_filter  = ...
    uint8(double(shadedImage) - (inverseShade_filter));

figure
imagesc(correctedImage_filter)

figure
xRow=300;
plot((1:columns), correctedImage_filter(xRow, :, 1), 'r-',...
    (1:columns), correctedImage_filter(xRow, :, 2), 'g--',...
    (1:columns), correctedImage_filter(xRow, :, 3), 'b-.',...
    'linewidth', 2);
grid on;
axis([1 columns 0 255])
xRow=400;
figure
plot((1:columns), correctedImage_filter(xRow, :, 1), 'r-',...
    (1:columns), correctedImage_filter(xRow, :, 2), 'g--',...
    (1:columns), correctedImage_filter(xRow, :, 3), 'b-.',...
    'linewidth', 2);
grid on;
axis([1 columns 0 255])
```

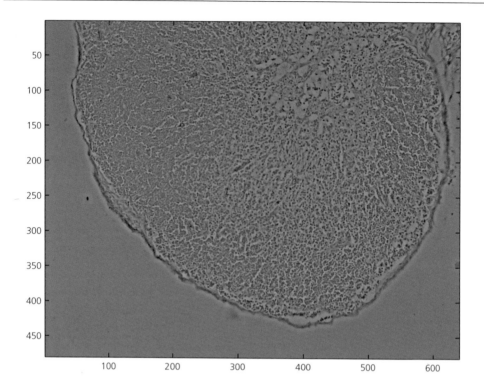

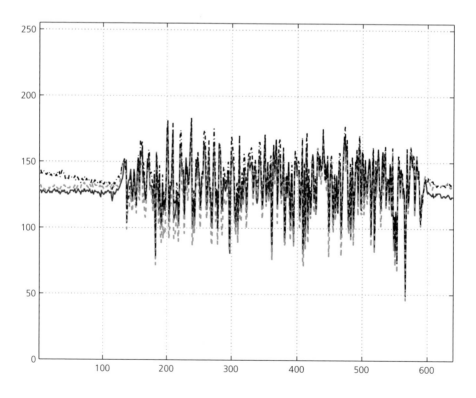

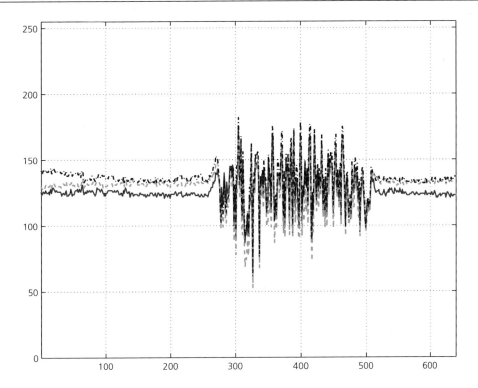

Both profiles show a very uniform intensity for the background regions. However, the image itself seems to have lost contrast; the darker cells seem to have lost their darker colours.

If the region with cells had been less dense, a better approximation could have been achieved. Similar approximations can be obtained by using morphological operators.

Morphological operators modify an image according to a predefined element or rule. For example, imagine a binary image with some foreground regions covered with ones and background denoted by zeros. We can "erode" that image by removing one or more pixels of the boundaries between the foreground and the background. Alternatively we can "dilate" the image by adding one or more pixels to the same boundaries. Interestingly, dilation and erosion are not reversible operations, as when we erode an image, a region may disappear, or when we dilate an image, inner holes in a region may be filled. Therefore, the combination of erosion and dilation depends on which is applied first: erosion followed by dilation is called "opening" and dilation followed by erosion is called "closing". Matlab provides commands to perform erosion ('imerode'), dilation ('imdilate'), closing ('imclose') and opening ('imopen') on images.

For our shading correction example, we can perform a closing on the image to estimate the shading component like this:

```
shadingComponent_close = imclose(shadedImage_grey,strel...
                    ('disk',9));
```

The overall effect of this operation is to "close" regions that are smaller than the structural element, but those areas that are larger than the element remain unchanged. For our present example, the structural element is a disk of radius 9. The overall effect will be to leave the background unchanged, and the regions with cells will be "closed". To visualise the effect, we can first display the estimated shading component:

```
figure
mesh(shadingComponent_close(1:4:end, 1:4:end, 1))
view(-140, 25)
axis tight
```

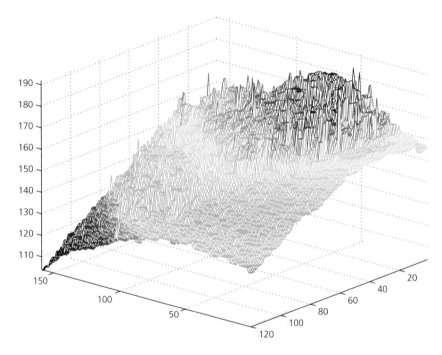

We can see now the effect of the closing. Unlike the previous estimation with the filter, the intensities in the area with cells are higher in intensity than before. The estimation of the background seems equally good as the previous case.

If we now observe the corrected image we can assess the suitability of the morphological operator to estimate the shading component.

```
inverseShade_close   = repmat(shadingComponent_close-...
                       mean(shadingComponent_close(:)),
                       [1 1 3]);

correctedImage_close = uint8(double(shadedImage) -...
                       (inverseShade_close));

figure
imagesc(correctedImage_close)
```

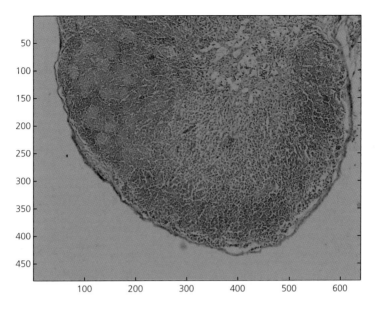

This new image shows that the background has been corrected and at the same time the contrast has not been affected. However, the region with cells does not appear to be uniform in intensity as there are some block artefacts. If we zoom in to the image and plot next to the estimated shading component, we can see the origin of these artefacts:

```
figure
subplot(121)
imagesc(correctedImage_close(100:250, 50:250, :))
subplot(122)
imagesc(shadingComponent_close(100:250, 50:250))
```

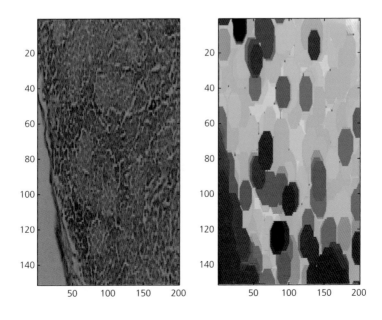

The shading component shows clearly that the origin of the block artefacts is implicit in the closing morphological operator that was used to estimate the shading component.

All the methodologies described so far have only partially solved the problem introduced by shading for this particular example. It is therefore important to mention that if the shading component that corrupted the original image had been flat and not curved, the plane would have provided a satisfactory result. If the cells have been arranged in a different way, sparsely spaced with background regions around them, then the filter or the closing could have provided satisfactory results. Therefore, the solution to many biomedical image problems is closely related to the nature of the data, i.e. the cells, and to the nature of the problem, i.e. the source of the shading.

As a final methodology in this recipe, we will describe another technique to estimate the shading component. We will not present all the steps, as it is a rather complex algorithm. The function used is 'shadingCorrection'. This is not a built-in function of Matlab and can be downloaded from the book's web site.

The algorithm of 'shadingCorrection' estimates the envelope that stretches over the data. The process of envelope estimation can be understood as the iterative stretching of a thin flexible surface under which (or over which) a series of objects (the data itself, cells in this case) are placed. Initially, the surface is identical to the signal intensity but after a series of stretches, the surface adapts to the peaks (or lowest points) of the objects, and intermediate values in between them. Therefore two envelopes can be generated, one below the data and one above. The envelope that is closest to the data, e.g. lowest difference, is selected as the shading component of the data. This algorithm is described in detail in Reyes-Aldasoro 2009, Electronics Letters.

The function 'shadingCorrection' receives as input argument the original image and returns the estimated shading component and the corrected image:

```
[correctedImage_envelope,shadingComponent_envelope]=...
    shadingCorrection(shadedImage);
```

Finally, we can display the shading component, the corrected image and the profile lines to compare with all the previous methodologies.

```
figure
mesh(shadingComponent_envelope(1:4:end, 1:4:end, 1))
view(-140, 25)
axis tight

figure
imagesc(correctedImage_envelope/255)

figure
xRow=300;
```

```
plot((1:columns), correctedImage_envelope(xRow, :, 1),...
    'r-',...
    (1:columns), correctedImage_envelope(xRow, :, 2),...
    'g--',...
    (1:columns), correctedImage_envelope(xRow, :, 3),...
    'b-.', 'linewidth', 2);
grid on;
axis([1 columns 0 255])
xRow=400;
figure
plot((1:columns), correctedImage_envelope(xRow, :, 1),...
    'r-',...
    (1:columns), correctedImage_envelope(xRow, :, 2),...
    'g--',...
    (1:columns), correctedImage_envelope(xRow, :, 3),...
    'b-.', 'linewidth', 2);
grid on;
axis([1 columns 0 255])
```

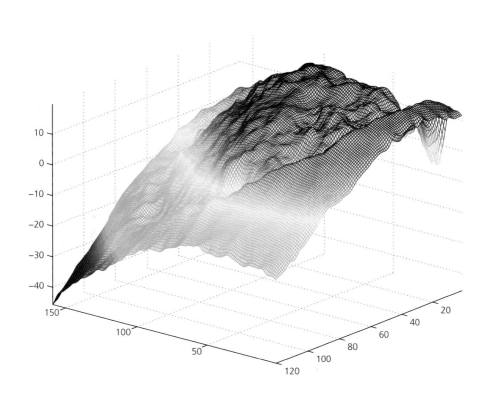

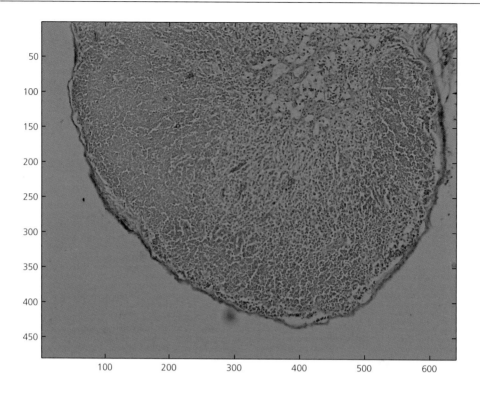

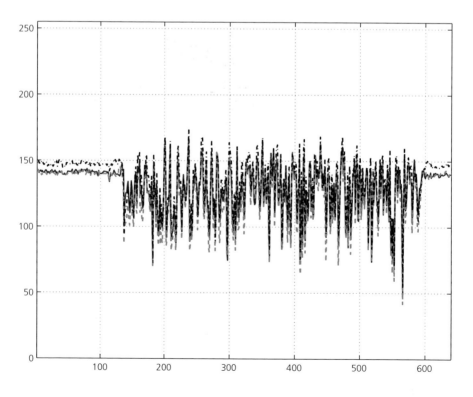

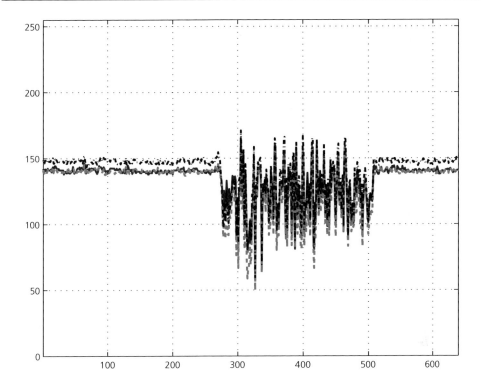

This algorithm has removed the shading of the image. The shading estimation is very close to the background in the areas without cells and follows the same trend in the areas with cells. The profiles show that the shading has been successfully removed.

To illustrate the importance of shading correction, we can now compare the segmentations that can be obtained by simple thresholding of the image, with and without shading correction. We will threshold the first channel of the images with a manually selected threshold of 128. The pixels above the threshold will be considered to be background, whilst the pixels below will be considered to be cells. We can segment and display in the same instruction like this:

```
figure
imagesc(double(shadedImage).*...
  (repmat(shadedImage(:, :, 1)<128, [1 1 3]))/255)

figure
imagesc(correctedImage_envelope.*...
  (repmat(correctedImage_envelope(:, :, 1)<128,...
  [1 1 3])/255))
```

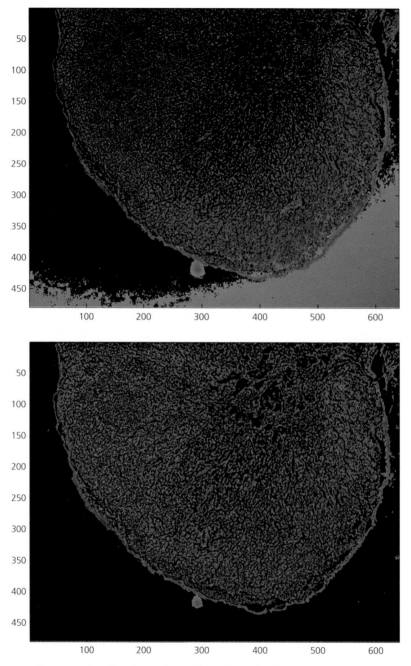

The two images clearly show the effect that shading can have; in the original shaded image, the background is divided in half as the darker region of the background is considered to be cells. Conversely, when the shading is removed, the background is completely removed with a simple thresholding operation. Furthermore, in the shaded image, some cell regions are not included in the segmentation (top centre).

This recipe has presented several techniques to correct shading on images, but also it has shown that simple techniques such as thresholding can be successfully applied to images of good quality.

6.4 Recipe 4: Mosaicking and Stitching

In several microscopical settings, there is a compromise between the magnification at which images are acquired and the regions of interest that are to be observed. In some cases, it is not possible to observe the whole region of interest at the magnification desired. Then, with the aid of a stage, either manual or motorised, the samples are moved so that images of different regions can be acquired.

With the development of motorised stages, the samples are moved precisely so that images can be easily aligned, but when precise motorised stages are not available, the displacements are performed manually, normally with some overlap between images in order to ensure that all areas are covered.

These images are later analysed collectively, or in other cases, they are manually aligned or "stitched" together, either manually or using software, either commercial or open source, to do so in a process referred to as mosaicking, aligning or stitching. In many cases, this process is not trivial and may be very time consuming.

In this recipe we will analyse some ways to align and stitch images together.

The simplest case is to merge images that have perfect alignment. In a case like this one, it is only necessary to combine matrices to form a larger matrix with all the smaller images.

As an example, we will align 18 'jpeg' images, which correspond to a larger sample that has been divided in three rows and six columns. The images have stored together inside a folder called 'sections'. To read the contents of the folder we can use the command 'dir' to display the contents of a directory.

```
dir sections/*.jpg
```

```
unshadedImage_1_1.jpg    unshadedImage_2_2.jpg    unshadedImage_3_3.jpg
unshadedImage_1_2.jpg    unshadedImage_2_3.jpg    unshadedImage_3_4.jpg
unshadedImage_1_3.jpg    unshadedImage_2_4.jpg    unshadedImage_3_5.jpg
unshadedImage_1_4.jpg    unshadedImage_2_5.jpg    unshadedImage_3_6.jpg
unshadedImage_1_5.jpg    unshadedImage_2_6.jpg
unshadedImage_1_6.jpg    unshadedImage_3_1.jpg
unshadedImage_2_1.jpg    unshadedImage_3_2.jpg
```

Notice that the names of the files include their corresponding row and column. We begin by reading the contents of the folder. The files could be read sequentially one by one, but in this case, since the file names include the row and the column where they will be placed, we can read them individually by using two 'for-loops', one corresponding to the rows and another corresponding to the columns.

First, we will read the images and display them in separate subplots to visualise each section separately. To do this, we will loop over the rows and columns and use the numbers to create the name of the file to read. The name is generated by a concatenation of strings; this is done with the command 'strcat', which stands for "string concatenate". We are concatenating five strings: (1) the first part of the name, which contains the folder ('sections') and the base name ('unshadedImage_'). Notice that the slash ("/") indicates a change of folder, (2) the number of the row; since this is a number (a double) we need to convert it to a string with the command 'num2str', which stands for "number to string", (3) an underscore "_", (4) the column, which again is converted to string and (5) the extension of the files ('.jpg'). It is important to highlight that a good naming convention for files can simplify analysis on cases like this one.

Once the name has been generated, this becomes the input argument for the command 'imread' to read the corresponding image. We then select a subplot of the figure with the current row and column and display the image in that location. The following code corresponds to the previous description.

```
figure
for counterRow=1:3
  for counterColumn=1:6
    sectionName  = strcat('sections/unshadedImage_', ...
        num2str(counterRow), '_', num2str...
        (counterColumn), '.jpg');
    currentSection = imread(sectionName);
    subplot(3, 6, (counterRow-1)*6+counterColumn)
    imagesc(currentSection)
  end
end
```

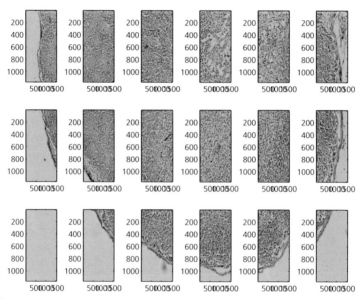

Notice how the text of the horizontal labels has overlapped and is unreadable. The overlap occurs because each image has three values or "ticks" over the axis. We can set the ticks of the axis easily, but first let's see what the value of these ticks are by "getting" the value from the current axis ('gca') like this:

```
get(gca, 'xtick')
```

```
    ans =
          500         1000         1500
```

We can see that the label is displaying three values of 500, 1000 and 1500. We can reduce from 3 to have only 2 ticks that are more spaced to avoid the overlap like this:

```
for counterRow=1:3
  for counterColumn=1:6
    subplot(3, 6, (counterRow-1)*6+counterColumn)
    set(gca,'xtick',[100 1500])
  end
end
```

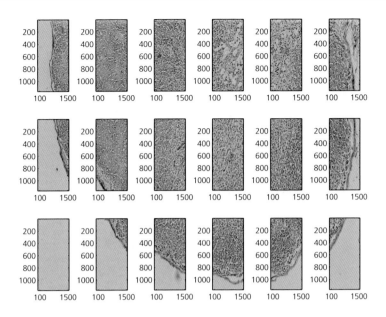

The basic technique to combine two separate matrices into a new one is to concatenate them by placing one after the other, either by extending the number of columns, that is '[A B]', or extending the rows, that is '[A;B]'. In the first case, it is necessary that matrices A and B have the same number of rows. In the second case, the matrices should have the same number of columns. If one matrix is an empty matrix (like 'A= [] ;'), then the combination of the two will be the same as the non-empty matrix. This is useful, as we do not need to initialise the matrices to be combined, nor do we need to know their dimensions.

For our case, we can create a matrix in which all the sections of a row will be appended as they are read. Then, this "row-matrices" will be appended into what will be the final complete image. We will do the same loops but instead of displaying the sections we will combine them.

```
combinedImage =[];
for counterRow=1:3
  combinedRow =[];
  for counterColumn=1:6
    sectionName   = strcat('sections/unshadedImage_',...
        num2str(counterRow), '_', num2str...
        (counterColumn), '.jpg');
    currentSection = imread(sectionName);
    combinedRow = [combinedRow currentSection];
  end
  combinedImage = [combinedImage;combinedRow];
end

figure
imagesc(combinedImage)
```

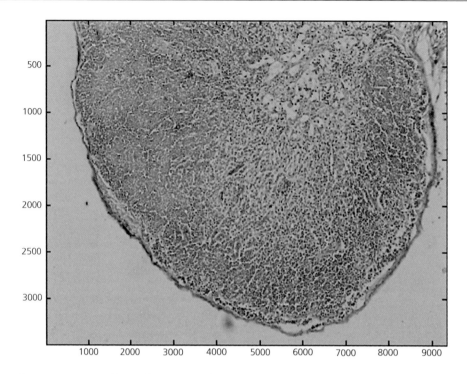

The 18 sections have been seamlessly arranged into a single image. We can compare the dimensions of an individual section against the complete image:

```
size(currentSection)
size(combinedImage)
```

```
    ans =

            1165          1556              3


    ans =

            3495          9336              3
```

The simplicity of the previous mosaicking relies on the fact that the sections were perfectly aligned to begin with; therefore it was only necessary to place them next to each other to create a combined image. Needless to say, this is not always the case.

We will now combine images that are not perfectly aligned. These images correspond to four quadrants that were acquired from a dorsal window chamber. The images are 'tiffs' and have been saved in a folder called 'quadrants'. Again, we can display the contents of the folder.

```
dir quadrants/*.tif
```

```
    Q11.tif  Q22.tif  Q33.tif  Q44.tif
```

Since we are dealing with only four images, it is not necessary to run a 'for-loop'. We can use four instructions to read each of them, like this:

```
Q1= imread('quadrants/Q11.tif');
Q2= imread('quadrants/Q22.tif');
Q3= imread('quadrants/Q33.tif');
Q4= imread('quadrants/Q44.tif');
```

All the images have the same size, and we will be using the dimensions later on, so we can calculate them with the command 'size'.

```
[rows, columns, channels]    = size(Q1);
```

As in most image-related problems, an initial observation of the data is important to understand the nature of the problem to solve.

```
figure
subplot(221)
imagesc(Q1)
subplot(222)
imagesc(Q2)
subplot(223)
imagesc(Q3)
subplot(224)
imagesc(Q4)
```

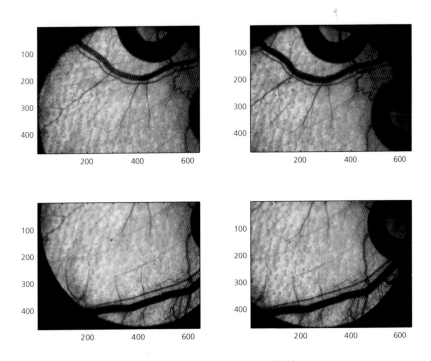

Images courtesy of Professor Gillian Tozer, The University of Sheffield, UK.

We can see that each image is covering a region of what is a single circular area. To verify that these four images do not match at the borders, we can display them together in the same way we combined the sections before:

```
figure
imagesc([Q1 Q2;Q3 Q4])
```

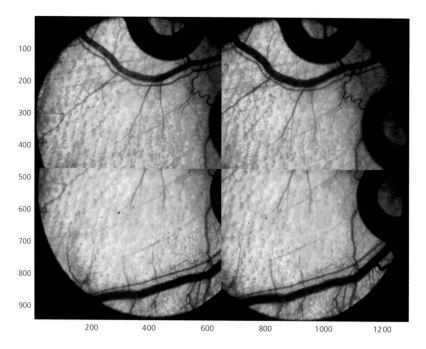

It is clear now that the images do not match and that there is considerable overlap between them. The problem is now one of finding the best alignment of the images and that can be done by shifting or displacing the images from their location, comparing them and then assessing the way they match each other. For this example, we can start with the top two images. First we will average them, without any shift, and then compare against the average of the images with some shift. To manually shift the images, we can discard the first 50 columns of one and the last 50 columns of the other and then average those images. We can visually compare the two results like this:

```
figure
imagesc(0.5*Q1+0.5*Q2)
title('Unshifted images', 'fontsize', 16)
```

```
figure
imagesc(0.5*Q1(:, 51:end, :)+0.5*Q2(:, 1:end-50, :))
title('Manually shifted images', 'fontsize', 16)
```

Unshifted images

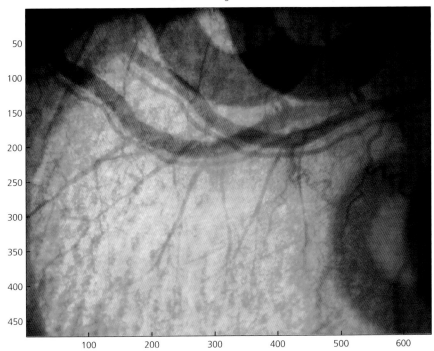

Manually shifted images

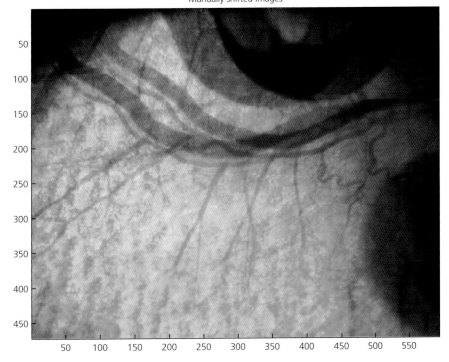

We can see that with the shifts, the vessels get closer to each other. We could continue shifting the images until we find a best match, however, this process has two clear deficiencies; first, it is laborious and time consuming to shift the images by manually selecting a certain displacement. Second, the measure of "best match" is a visual measure. If the images are slightly blurred, it would be very subjective to determine a best match.

Clearly, we need to determine a quantitative measure of similarity between images, which will then be used in an automated shifting process. Both of these requirements of measuring and shifting are obtained through a mathematical process called "cross-correlation". Cross-correlation is a process in which two signals are multiplied and the product is summed. One of the two signals is continuously shifted, in theory from minus infinity to plus infinity, in practice from one edge of the image to the other edge. The output of this operation is a single value, which will vary according to the displacement or shift of the signal. The value of the cross-correlation is proportional to the similarity between the two signals, i.e. if a signal is multiplied by itself and then summed, the value would be the sum of the squared elements. If the product of the signals has a similar number of negative and positive values with similar magnitudes, the sum will tend to 0. If the signals are periodical, like a square wave or a sinusoid, then the correlation will also be periodical, as signal will repeat itself after some pixels. In our example, the maximum value of the cross-correlation will indicate the shift that will best match the two images.

The result of the cross-correlation is a matrix with dimensions nearly double of the original images; it is actually of the size [2*rows-1, 2*columns-1]. If the two images were identical, the maximum value of the cross-correlation would be at the centre of the matrix, at [rows, columns], therefore we can decide that the best match would be determined by the position at which we can find the maximum value.

Matlab has a command to calculate the cross-correlation between 1D signals ('xcorr') and 2D signals, i.e. images, ('xcorr2'). However, before using cross-correlation we need one pre-processing step to obtain better results. Instead of obtaining the cross-correlation between the original images, we can obtain more accurate results if the cross-correlation is calculated between images that have more salient characteristics. To do this, we can detect the edges of the images, and thus the edges will provide a better match than the raw intensities of the original images.

Finding edges in an image is a very important and basic step in image processing. The basic step to find an edge in an image is to look for regions where the intensity changes abruptly. To do that, we need an operation that highlights those changes, whilst it does not highlight regions of uniform or constant value. The operation in question could be a difference, which is the basis of the high-pass filters, which are covered in more detail in the section for Scratch Wound Assays. Matlab provides the command 'edge' that is specifically designed to find edges in images. The command 'edge' can be used with several edge-detecting algorithms; one that is widely used is the "Canny" algorithm, which uses several post-processing steps to detect the most salient edges at different scales. We will use the command 'edge' with the same parameters to find the edges of the four quadrants of our example, but it is highly recommended that you test this command with different parameters and observe the impact that the parameters and algorithms can have on the edges.

```
[Q1BW]=double(edge(Q1(:, :, 1), 'canny', [], 3));
[Q2BW]=double(edge(Q2(:, :, 1), 'canny', [], 3));
[Q3BW]=double(edge(Q3(:, :, 1), 'canny', [], 3));
[Q4BW]=double(edge(Q4(:, :, 1), 'canny', [], 3));
```

```
figure
subplot(221)
imagesc(Q1BW)
subplot(222)
imagesc(Q2BW)
subplot(223)
imagesc(Q3BW)
subplot(224)
imagesc(Q4BW)

colormap (gray)
```

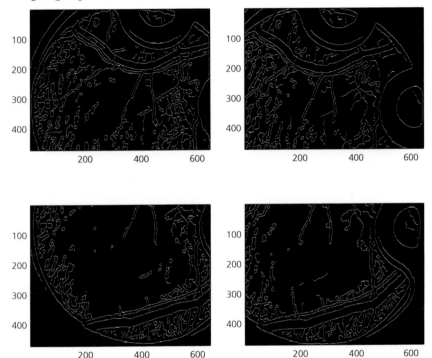

We can now see the edges of the four quadrants that we wish to align. It is over these binary images that we will perform the cross-correlation. An interesting experiment is to find the cross-correlation of one image with itself. This is called an "auto-correlation" and will be useful when compared against the cross-correlation of that image with another image.

We can calculate the cross-correlation between the first two images, and the auto-correlation of the first image with the command 'xcorr2' like this:

```
crossCorrelation_1_2                         = xcorr2(Q1BW, Q2BW);
crossCorrelation_1_1                         = xcorr2(Q1BW, Q1BW);
```

We can display the results of the two correlations in a single figure. We will use two meshes to compare the position of the maximum values:

```
figure
mesh(crossCorrelation_1_2/max(crossCorrelation_1_2(:)),...
    'edgecolor', 'b', 'linewidth', 2)
```

```
hold on
mesh(crossCorrelation_1_1/max(crossCorrelation_1_1(:)),...
     'edgecolor', 'r', 'linewidth', 2)
axis tight
view(-5, 20)
```

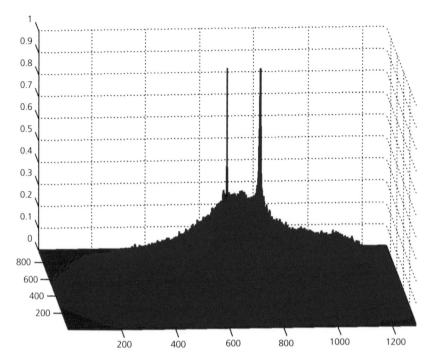

From the figure we can see that both correlations have a very clear and narrow maximum peak, which is not located in the same coordinates. It is this difference in locations that we are seeking, as we will use that to shift the images for the alignment. We can now calculate the position of the maximum value of each case. To do this, we will first use the command 'max' to find the maximum value; next we compare all the values of the cross-correlation against this maximum. The output of that comparison will produce a 1 for all pixels (it may be 1 or more) that have the maximum value and a 0 elsewhere. Finally, we use the command 'find' to find the location of the non-zero pixels of the matrices. We can do the three steps in one single line of instructions like this:

```
[rr_1_1, cc_1_1] = find(crossCorrelation_1_1==...
                   max(crossCorrelation_1_1(:)));
[rr_1_2, cc_1_2] = find(crossCorrelation_1_2==...
                   max(crossCorrelation_1_2(:)));
```

The outputs of the command 'find' are the rows and columns in which the non-zero elements of the matrix are located. Thus we now need to compare those values to find the displacement, in rows and columns, between the peaks. To do that, we simply subtract the rows and columns of the cross-correlation from the auto-correlation.

```
shift_Rows_1_2              = rr_1_2 - rr_1_1;
shift_Cols_1_2              = cc_1_2 - cc_1_1;
```

We can display the results with the command 'disp':

```
disp([ rr_1_1 cc_1_1; rr_1_2 cc_1_2;...
   shift_Rows_1_2 shift_Cols_1_2])
```

```
   472    644
   463    766
    -9    122
```

It should not be surprising that the peak of the autocorrelation is at the centre of the matrix, and the coordinates coincide with the original dimensions of the image, in this case [472 644]. It is interesting to notice that the peak of the cross-correlation between images 1 and 2 is displaced by 9 rows and 122 columns. We can repeat the calculation for images 3 and 4:

```
crossCorrelation_3_4      = xcorr2(Q3BW, Q4BW);
crossCorrelation_3_3      = xcorr2(Q3BW, Q3BW);
[rr_3_4, cc_3_4]          = find(crossCorrelation_3_4==...
                               max(crossCorrelation_3_4(:)));
[rr_3_3, cc_3_3]          = find(crossCorrelation_3_3==...
                               max(crossCorrelation_3_3(:)));

shift_Rows_3_4            = rr_3_4 - rr_3_3;
shift_Cols_3_4            = cc_3_4 - cc_3_3;

disp([rr_3_3 cc_3_3; rr_3_4 cc_3_4;...
   shift_Rows_3_4 shift_Cols_3_4])
```

```
   472    644
   463    785
    -9    141
```

The shift between images 3 and 4 is the same in the rows but slightly higher in the columns.

We can now align the first two images. We could discard rows and columns as we did previously, but there is a simpler way to arrange the matrices as discarding rows and columns may be complicated; we need to be careful with the dimensions of the cropped images. Matlab provides a command ('circshift') that shifts the rows and columns of a matrix in a circular way, that is, column 3 can be placed in column two, column two in column one and column one becomes the last column of the matrix when we shift by one to the left. Since we have calculated the shifts in rows and columns previously, we can calculate the shifted versions of the quadrants easily:

```
Q2_shifted      = circshift(Q2,...
                     [shift_Rows_1_2 shift_Cols_1_2]);
Q4_shifted      = circshift(Q4,...
                     [shift_Rows_3_4 shift_Cols_3_4]);
```

Now we can merge the shifted quadrants with their matching images by concatenating two matrices into a new matrix like this:

```
Q_1_2           = [Q1 Q2_shifted];
Q_3_4           = [Q3 Q4_shifted];
```

```
figure
imagesc(Q_1_2)
```

```
figure
imagesc(Q_3_4)
```

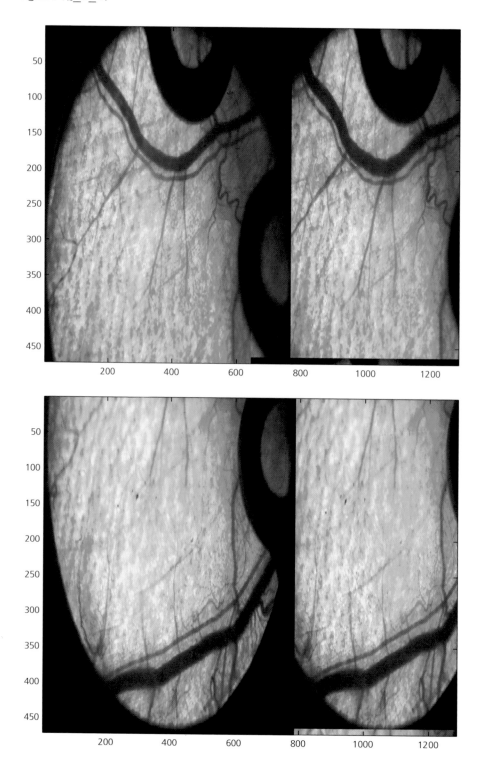

The merged images present a curious view; the images have been merged and present a more complete version of the original view. However, due to the circular shift a repeated segment of the image is present at the right hand side and a very small segment at the bottom. Since the shift between images 1 and 2 is different from the shift between 3 and 4, we face a compromise before we can merge vertically between the previously merged segments. One option is to follow the smallest image and guarantee that all pixels in a final image come from one of the quadrants; the other option is to follow the largest image and pad on the edges with blank pixels. In this example, we will follow the first approach; therefore we will now crop each result before looking for the match between Q_1_2 and Q_3_4. Fortunately, we know how much we shifted the images, so we can use these values as a guide to calculate the dimensions to crop both images:

```
column_crop       = min(shift_Cols_1_2, shift_Cols_3_4);
row_crop          = min(shift_Rows_1_2, shift_Rows_3_4);

Q_1_2_cropped     = Q_1_2(1:rows +row_crop,...
                    1:columns+column_crop,:);
Q_3_4_cropped     = Q_3_4(1:rows +row_crop,...
                    1:columns+column_crop,:);
```

```
[rows_12, columns_12, channels_12] = size(Q_1_2_cropped);
```

We now have two images to match; in this case, the stitching must be done vertically and not horizontally as it was done before. We can visualise the process so far, by placing one intermediate result above the other one:

```
figure
imagesc([Q_1_2_cropped;Q_3_4_cropped])
```

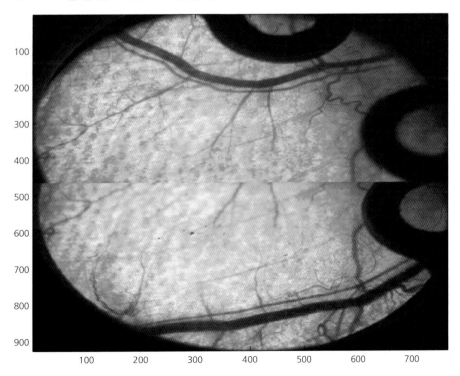

Before we perform the cross-correlation between the top two images and the bottom ones, we need to obtain the edges as done previously:

```
[Q12BW]=double(edge(Q_1_2_cropped(:, :, 1), 'canny', [], 3));
[Q34BW]=double(edge(Q_3_4_cropped(:, :, 1), 'canny', [], 3));
```

Now we can calculate the cross-correlations, the positions of their maximum values and the corresponding shifts:

```
crossCorrelation_12_34    = xcorr2(Q12BW, Q34BW);
crossCorrelation_12_12    = xcorr2(Q12BW, Q12BW);
[rr_12_34, cc_12_34]      = find(crossCorrelation_12_34==...
                            max(crossCorrelation_12_34(:)));
[rr_12_12, cc_12_12]      = find(crossCorrelation_12_12==...
                            max(crossCorrelation_12_12(:)));

shift_Rows_12_34          = rr_12_34 - rr_12_12;
shift_Cols_12_34          = cc_12_34 - cc_12_12;
```

We will use the values calculated to perform the circular shift of the bottom two images, 3 and 4, which will later be stitched with the top two images, 1 and 2.

```
Q34_shifted               = circshift(Q_3_4_cropped, ...
                            [shift_Rows_12_34 shift_Cols_12_34]);
```

Now we can merge the shifted quadrants with their matching images and visualise the results so far:

```
Q_1234                    = [Q_1_2_cropped;Q34_shifted];
figure
imagesc(Q_1234)
```

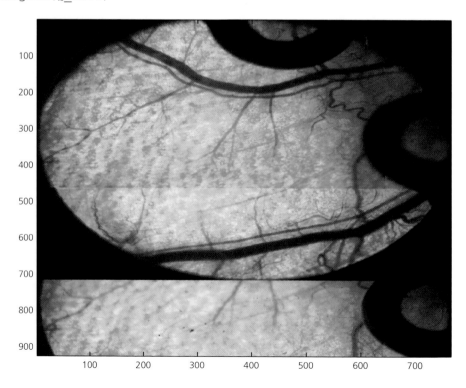

Again, it is necessary to crop the image to remove the repeated section in the bottom of the merged image:

```
Q_1234_final          = Q_1234(1:rows_12+shift_Rows_12_34,...
                              1:columns_12+shift_Cols_12_34, :);
figure
imagesc(Q_1234_final)
```

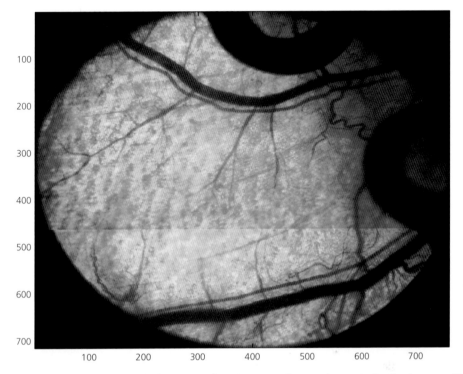

We have now stitched four overlapping quadrants into a single image. The stitching is quite good, however, the edges between the former quadrants are visible. This is not due to a poor matching but rather a result of different initial conditions of the original images. The intensity is not the same at the top than at the bottom, and there is a small grey vertical line. With a few pre-processing steps, these details could be corrected, for instance, cropping a few rows and columns of each images before they are stitched, or modifying the brightness and contrast so that the images would have the same initial conditions.

These steps may seem involved to be repeated each time that images need to be stitched, but all of them can easily be saved as one function; one that takes as input arguments four quadrant images, or pairs if so required, and output the one single stitched image. In this way, merging images would require only one line of code.

6.5 Recipe 5: Pixel Intensity and Histograms in Immunohistochemistry

The process of detecting proteins in tissue samples by using antibodies that bind to specific antigens is one of the most popular techniques of tissue analysis in the Life Sciences and is known as 'immunohistochemistry', sometimes abbreviated as

"IHC". The IHC applications range from detection of abnormal cells to the observation of distribution of markers in biological tissue.

Immunohistochemistry can be combined with the use of fluorophores, and it is sometimes referred to as 'immunofluorescence'. However, in this section we will concentrate on staining methods that do not use fluorescence, as the distribution of colours with fluorescence is normally associated with the primary colours red, green and blue, whilst when fluorescence is not used, the reporter molecules yield an intense colour that can be observed under a traditional light microscope.

To begin with the analysis let's first read an image. In the same way that we have done in other sections, we can read a very good quality image from Wikipedia (uploaded to Wikimedia Commons by 'Nephron' 10 April 2011):

```
immunohisto_CD31 = imread(['http://upload.wikimedia.org/',...
   'wikipedia/commons/4/4c/Epithelioid_angiosarcoma_-_',...
   'CD31_-_intermed_mag.jpg']);
imagesc(immunohisto_CD31)
```

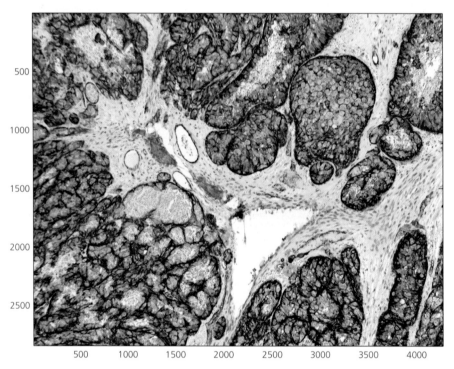

A first observation is that where there is no tissue present in the image, the pixels appear with a white colour, if the image is of good quality, but can vary in the range of greys with lower quality images. The implication is that, unlike fluorescent images, there will be a component of these bright tones in the three channels that form the image. To illustrate this, we can display separately the three channels linked to the primary colours red, green and blue, with the following commands:

```
figure
subplot(131)
imagesc(immunohisto_CD31(:, :, 1))
subplot(132)
imagesc(immunohisto_CD31(:, :, 2))
```

```
subplot(133)
imagesc(immunohisto_CD31(:, :, 3))
```

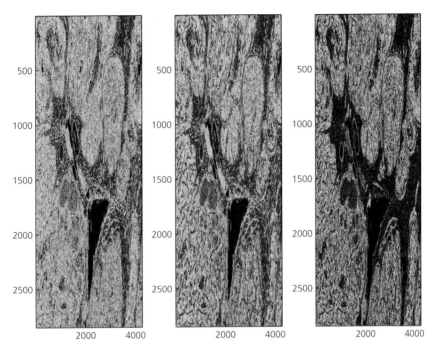

The previous figure is not very clear, as the `'jet'` colour map is unsuitable for this image. It would be better to use a grey-scale colour map:

```
colormap(gray)
```

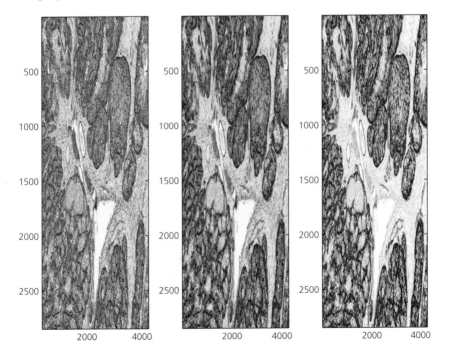

In grey scale, the image is easier to compare to the original in colour, but still we cannot identify structures that are present in one channel and absent in another. The colours present in the image, brown and light blue-purple, are both a mixture of red, green and blue.

We can now see that it is not possible to separate cells or regions as easily as it is the case with fluorescence and therefore we will need more sophisticated techniques. To introduce a powerful analysis technique, it is first necessary to introduce first the concept of colour spaces.

In subsequent recipes we will investigate the colour of the pixels, but first we need to learn about some image processing techniques that rely solely on the intensity of the pixels of an image; that is, how bright or dark they are, and the relationship to other pixels around them. To begin, let's first take one image of a bone marrow smear that has been stained with Wright's stain (uploaded to Wikimedia Commons by 'Dybdal' 19 May 2007):

```
immunohisto_Wright = imread(['http://upload.wikimedia.org/',...
    'wikipedia/commons/0/0e/Acute_leukemia-ALL.jpg']);
```

As this image is a colour image, we will convert it from the red, green, blue format to grey scale with the command 'rgb2gray'. In addition, we will convert the result from the integer format 'uint8' to 'double' as these are easier to manipulate later on.

```
immunohisto_Wright_gray = ...
    double(rgb2gray(immunohisto_Wright));
figure
subplot(121)
imagesc(immunohisto_Wright)
subplot(122)
imagesc(immunohisto_Wright_gray)
colormap(gray)
```

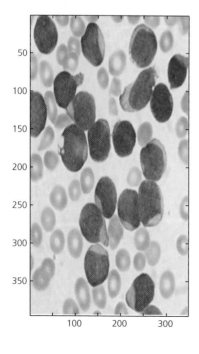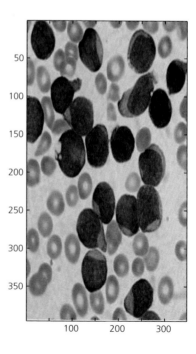

A visual inspection quickly shows that there are three ranges of grey levels present in the image. First there is a bright background that tends towards white. Second, there is an intermediate range of greys that correspond to the smaller round cells and finally a darker region with large cells. Histograms are mathematical techniques that can inform and quantify the visual inspection just performed.

Intensity histograms are well-known measurements of the occurrence or relative frequency of values and, in the case of images, these can measure intensities or grey levels. Histograms are often the basis of threshold-based segmentation methods as they can show where the optimal separation of two classes is, when these classes are clearly distributed into two regions of the histogram. The command 'hist' calculates the histogram of an image in the following way:

```
figure
[intensityOccurrence, intensityAxis]    = ...
    hist(immunohisto_Wright_gray(:), 40);
bar(intensityAxis, intensityOccurrence)
```

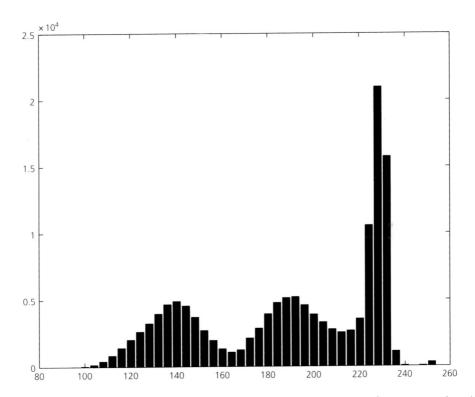

The function 'hist' counts the number of elements that occur in 40 equally spaced slots or "bins" and stores them in the first output variable (i.e. 'intensityOccurrence'). The centre of the bins is recorded in the second output parameter (i.e. 'intensityAxis'). What the histogram is doing is counting all the elements of the input (intensity in this case) that fall within the "edges" or limits of a bin. In the previous example we had 40 bins with edges equally distributed from 0 to 255.

Notice that the number of occurrences is quite high, and these numbers depend on the frequency of a given intensity as well as the size of the image. Histograms are also used as a measurement of the frequency or 'relative' occurrence, that is, as a per cent of the total. To display that, we need to divide the absolute occurrences by the total number of pixels to obtain the ratios:

```
relIntensityOccurrence = intensityOccurrence/...
                    sum(intensityOccurrence(:));
```

Another useful way to analyse the occurrences is to display the cumulative distributions, that is, show how the values accumulate along the x-axis until they reach the maximum or 1. To do this, we sum the occurrences in a cumulative way with the command 'cumsum'. The cumulative histogram is useful if we want to see how many pixels are below a certain level, or conversely to find the intensity at which a certain proportion of the image belongs.

```
figure
subplot(211)
bar(intensityAxis, relIntensityOccurrence)
axis ([0 255 0 0.16]); grid on
ylabel('Relative Occurrences')
subplot(212)
plot(intensityAxis, cumsum(relIntensityOccurrence),...
     'linewidth', 2)
axis ([0 255 0 1.05]); grid on
ylabel('Cummulative Occurrences')
xlabel('Pixel Intensity')
```

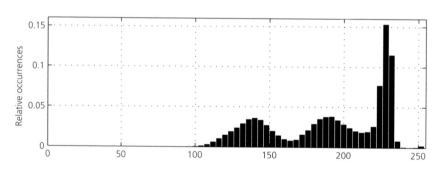

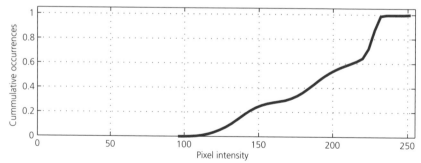

As we expected, the histogram is showing three "humps" or "peaks" corresponding to the three regions previously described. In addition, we can see that the valleys in between these hills, or regions of minima and maxima as they are technically termed, can indicate where we should aim to partition the image. We could measure these points manually, but Matlab has functions to help with this. The function 'imregionalmin' is designed to find regions of minima in images and can also be used in 1D matrices. The output of 'imregionalmin' is a matrix with a one where there is a value considered as minima and zeros otherwise. To find the exact position of these minimum locations, we can address the matrix of the intensity axis (Chapter 1 describes addressing matrices) in the following way:

```
intensityMinima      = imregionalmin(intensityOccurrence);
minimalThresholds    = round(intensityAxis(intensityMinima))
```

```
minimalThresholds =

   96    164    212    244
```

We have as a result four values: 96, 164, 212 and 244. The ones that are interesting in our case are the two middle ones, which we can now use directly to do a segmentation of the image into three classes or groups, two cells and the background:

```
image_background  = (immunohisto_Wright_gray>...
                     minimalThresholds(3));
dark_cells        = (immunohisto_Wright_gray<...
                     minimalThresholds(2));
bright_cells      = (immunohisto_Wright_gray<...
                     minimalThresholds(3)).*...
                     (immunohisto_Wright_gray>...
                     minimalThresholds(2));

figure
subplot(131)
imagesc(dark_cells.*immunohisto_Wright_gray)
caxis([0 255])


subplot(132)
imagesc(bright_cells.*immunohisto_Wright_gray)
caxis([0 255])


subplot(133)
imagesc(image_background.*immunohisto_Wright_gray)
caxis([0 255])
colormap(gray)
```

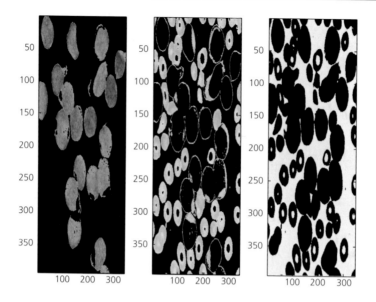

Although the initial segmentation is rather good, we can clearly see that there are problems: (1) there are some pixels inside the dark cells that were segmented as brighter cells, (2) there are some rings around the dark cells that were also segmented as brighter cells and (3) some pixels of the background were segmented incorrectly. To illustrate what is going on, we will use one technique that has been explored in more detail in other recipes, namely, "image dilation". Image dilation is one of several morphological operations that are easily performed in Matlab.

Image dilation consists of increasing the size of a given region or class, by adding pixels to the boundary of the region. The pixels that are added can follow different shapes like squares, disks or lines, which are sometimes called "kernels" or "structural elements". For the example we will use a very simple dilation of the previously segmented classes with a 3×3 square region. The command 'imdilate' is used to perform image dilation:

```
dilationSize        = 3;
dilationElement     = ones(dilationSize);

image_background_2  = imdilate(image_background, ...
                         dilationElement);
dark_cells_2        = imdilate(dark_cells, dilationElement);
bright_cells_2      = imdilate(bright_cells, dilationElement);
```

The regions that we have created with the dilation will overlap slightly, and we can see that some of the pixels that were not included previously as background or dark cells have been correctly assigned, and also the rings are now thicker lines.

```
figure
subplot(131)
imagesc(dark_cells_2.*immunohisto_Wright_gray)
caxis([0 255])
```

```
subplot(132)
imagesc(bright_cells_2.*immunohisto_Wright_gray)
caxis([0 255])

subplot(133)
imagesc(image_background_2.*immunohisto_Wright_gray)
caxis([0 255])
colormap(gray)
```

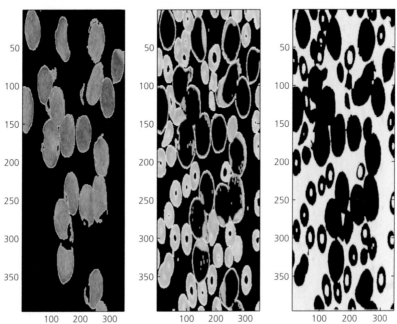

Now, we can obtain the histograms of each of the dilated regions. To do this, we need to multiply the segmentation by the original image. We can also define the region of intensity where we want the histogram to be calculated, not just a number of bins. We do this by defining the second parameter to the command 'hist'. In this case, we are interested in the region between 90 and 255.

```
[intensityDarkCells,intensityAxisDarkCells]       = ...
    hist(dark_cells_2(:).*immunohisto_Wright_gray(:),...
    [90:255]);
[intensityBrightCells,intensityAxisBrightCells] =...
    hist(bright_cells_2(:).*immunohisto_Wright_gray(:),...
    [90:255]);
[intensityBackground,intensityAxisBackground]     =...
    hist(image_background_2(:).*immunohisto_Wright_gray(:),...
    [90:255]);
```

All the pixels that are not part of the segmentation (i.e. the black pixels in each of the previous images) will be counted into the first bin of the histogram. Therefore, we will discard this first bin in the following plot by addressing from the second element of the matrix.

```
figure
hold on
plot(intensityAxisDarkCells(2:end),   intensityDarkCells...
     (2:end),'b')
plot(intensityAxisBrightCells(2:end),intensityBrightCells...
     (2:end),'k')
plot(intensityAxisBackground(2:end),  intensityBackground...
     (2:end),'r')

grid on;axis tight
legend('Dark Cells','Brighter Cells','Background','Location',...
       'NorthWest')
```

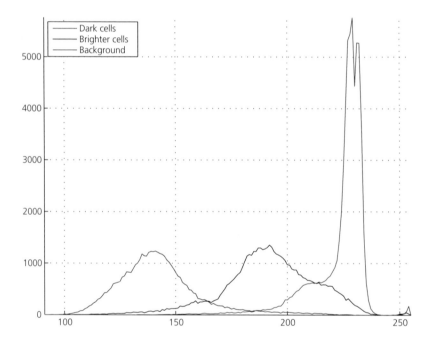

The three individual histograms now show how the pixels distribute their intensities, and we can see that the tails of each class spread along the axis of the intensities well into the regions that we had considered as belonging to the other classes. When we choose a threshold, we do so trying to minimise the number of pixels that will be incorrectly assigned to a class. In some cases, but not always, the lowest point in a histogram will help achieve this minimum error. In cases where one class is larger than another one, the thresholds may need to be shifted up or down to find the optimal solution. There are statistical techniques that can minimise the error based on certain assumptions, like a Gaussian distribution of the classes, but these are outside the scope of this book.

6.6 Recipe 6: Hue-Saturation-Value

Perhaps the most common colour space in use is the RGB, with red, green and blue channels. However, there are other important colour spaces used in different applications. For example, analogue colour television used to use a colour space that combined a grey-scale channel for the traditional "black and white" receivers and the chrominance or colours for the colour receivers. This space is called YUV for luminance and chrominance. In biomedical image analysis a common colour space is the Hue Saturation and Value (HSV). The HSV space describes perceptual colour relationships related to the artistic ideas of hue tint and shade, which provide a more natural description of colour than additive or subtractive colour components. Hue is a property related to the wavelength of the colour arranged in a circle around the Value axis and has values between 0 and 1, but can be interpreted as an angle between 0 and 360 degrees. As a reference, red corresponds approximately to 0, yellow to 0.15 (60 degrees), green to 0.33 (120 degrees), cyan to 0.5 (180 degrees), blue to 0.66 (240 degrees) and magenta to 0.87 (315 degrees). Saturation is a measure of the intensity or the purity of the hue so that a low saturated pixel will be white or grey, and a high saturated pixel will show its corresponding hue. The third channel, value, is a measure of the darkness or brightness. According to the way this last channel is calculated, it is sometimes called intensity, lightness, or brightness, and the corresponding colour spaces are called HSI, HSL or HSB.

The easiest way to understand the HSV colour space is to display it. To do this, we can generate an image that spans all the combinations of hue and saturation, and converts them from HSV to the familiar RGB colour space using the command 'hsv2rgb'. We will set two values, one at the maximum (i.e. 1 which corresponds to brightest) and one at 0.75. Each channel will vary from 0 to 1 in 50 steps:

```
numberSteps            = 50;
hueRange               = linspace(0, 1, numberSteps);
saturationRange        = linspace(0, 1, numberSteps);

[xData, yData]         = meshgrid(hueRange, saturationRange);
zData1                 = ones(numberSteps);
zData2                 = 0.75*ones(numberSteps);

image1_HSV(:, :, 1)    = xData;
image1_HSV(:, :, 2)    = yData;
image1_HSV(:, :, 3)    = zData1;
image1_RGB             = hsv2rgb(image1_HSV);

image2_HSV(:, :, 1)    = xData;
image2_HSV(:, :, 2)    = yData;
image2_HSV(:, :, 3)    = zData2;
image2_RGB             = hsv2rgb(image2_HSV);
```

Since we have two planes for each of the values, we will display them one on top of the other on the corresponding locations of the value axis.

```
figure
hold on
surf1 = surf(xData, yData, zData1, image1_RGB, 'EdgeColor',...
              'none');
surf2 = surf(xData, yData, zData2, image2_RGB, 'EdgeColor',...
              'none');
xlabel('Hue', 'fontsize', 16)
ylabel('Saturation', 'fontsize', 16)
zlabel('Value', 'fontsize', 16)
view(160, 30)
```

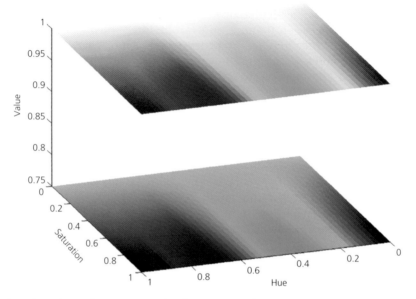

Notice that the colours when the hue is 0 and 1 are very similar. This is due to the circular property previously mentioned. Each of these planes is called a "loci of constant value".

Matlab can convert an image from the RGB space to an image in the HSV space with the command 'rgb2hsv'. We can read the same image as before and then convert it to HSV in the following way:

```
immunohisto_CD31 = imread(['http://upload.wikimedia.org/',...
  'wikipedia/commons/4/4c/Epithelioid_angiosarcoma_-_',...
  'CD31_-_intermed_mag.jpg']);
immunohisto_CD31_hsv     = rgb2hsv(immunohisto_CD31);
```

The size of the image will be useful later, so we can calculate the number of rows and columns with 'size':

```
[rows, columns, levels]  = size(immunohisto_CD31_hsv)

   rows =

      2848
```

```
columns =

        4272


levels =

        3
```

The two images, in RGB and HSV, will have the same dimensions in rows, columns and levels. To explore the HSV space we can first visualise the HSV individual channels together with the original image.

```
figure
subplot(141)
imagesc(immunohisto_CD31_hsv(:, :, 1))
title('Hue channel')
subplot(142)
imagesc(immunohisto_CD31_hsv(:, :, 2))
title('Saturation channel')
subplot(143)
imagesc(immunohisto_CD31_hsv(:, :, 3))
title('Value channel')
subplot(144)
imagesc(immunohisto_CD31)
title('Original Image')
```

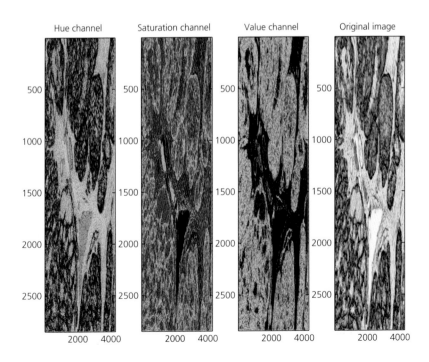

From a first glance, the image in the HSV space may be confusing, especially the image corresponding to the hue. Let's first analyse the saturation and the value components. This is easier if we change the colormap to grey (remember, this is different from the colour spaces; it is just how the values are displayed in the screen):

```
colormap(gray)
```

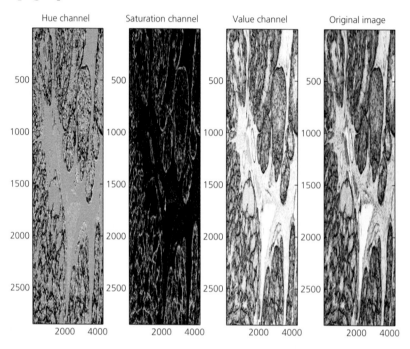

The areas where saturation is low (i.e. dark) is where the colours are very close to white or grey, i.e. there is no pure colour. The areas where the saturation is high (i.e. bright) is where the colours are purest or highly saturated. The value corresponds directly to the brightness or darkness of the original image.

We can now use thresholds to see the spatial distribution of the HSV channels: the lower range (0–0.2) in the hue will correspond to the region red to yellow; we will also select the highly saturated (>0.5) and low values (<0.5) pixels:

```
subplot(141)
imagesc(immunohisto_CD31_hsv(:, :, 1)<0.2)

title(strcat('Low Hues =', ...
    num2str(sum(sum(immunohisto_CD31_hsv(:, :, 1)<0.2))...
    /rows/columns, 2), ' %'))
subplot(142)
imagesc(immunohisto_CD31_hsv(:, :, 2)>0.5)

title(strcat('High Saturation =',...
    num2str(sum(sum(immunohisto_CD31_hsv(:, :, 2)>0.5))...
    /rows/columns, 2), ' %'))
subplot(143)
imagesc(immunohisto_CD31_hsv(:, :, 3)<0.5)
title(strcat('Low Values =',...
```

```
        num2str(sum(sum(immunohisto_CD31_hsv(:, :, 3)<0.5)))...
        /rows/columns, 2), ' %'))
subplot(144)
imagesc(immunohisto_CD31)
title('Original Image')
```

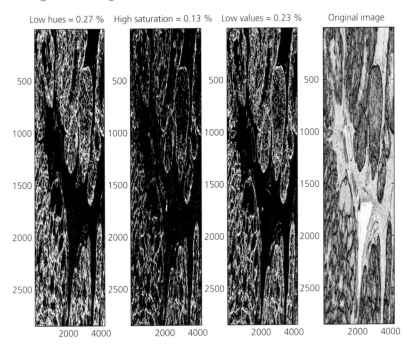

The figure should be clearer now, as the thresholds we have selected correspond roughly to the dark-brown pixels of the original image. It also helps to visualise that the majority of the pixels have low saturation and high values, that is, the image is bright and tends to white-grey colours.

To visualise the colour decomposition it is convenient to save each channel in a separate matrix, as we will want to manipulate these later:

```
currentHue          = immunohisto_CD31_hsv(:, :, 1);
currentSaturation   = immunohisto_CD31_hsv(:, :, 2);
currentValue        = immunohisto_CD31_hsv(:, :, 3);
```

Since the current image is quite large, we can subsample it, that is, take a reduced number of the pixels for analysis and rearrange them into a column position, as we will exploit this later on:

```
subSampleStep       = 2000;

sampledHue          = currentHue(1:subSampleStep:end)';
sampledSaturation   = currentSaturation(1:subSampleStep:end)';
sampledValue        = currentValue(1:subSampleStep:end)';
```

For the next display, we will use the function 'scatter3' previously introduced. This function takes as input parameters an x, y, z position, which in this case will be the hue, saturation and value for the sampled pixels, and the colour will be obtained from the original image. Therefore we will proceed in the same way, saving the RGB

channels in separate matrices and sub-sampling them. Additionally we will convert
the values to 'double' as they were originally 'uint8'.

```
currentRed          = immunohisto_CD31(:, :, 1);
currentGreen        = immunohisto_CD31(:, :, 2);
currentBlue         = immunohisto_CD31(:, :, 3);

sampledRed          = double(currentRed(1:subSampleStep:end)');
sampledGreen        = double(currentGreen(1:subSampleStep:end)');
sampledBlue         = double(currentBlue(1:subSampleStep:end)');
```

With the proper matrices, it is now time to display the data of the sampled pixels,
each with their original colour, positioned according to their corresponding hue,
saturation and value coordinates. We will use the command 'scatter3', which was
introduced in Chapter 3 colour.

```
figure
scatter3(sampledHue, sampledSaturation, sampledValue, 20,...
    [sampledRed, sampledGreen, sampledBlue]/255, 'filled');
rotate3d on
view(-18, 66)
xlabel('Hue')
ylabel('Saturation')
zlabel('Value')
```

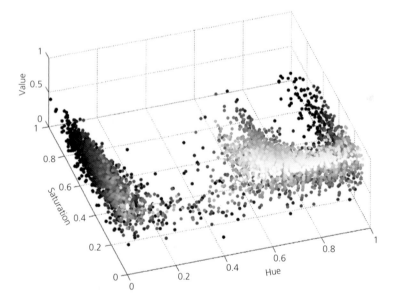

The command 'scatter3' has used three input arguments to position a sphere
of a fixed size (20), and the colour of that sphere is given by the fourth argument,
which is taken from the sample colours.

The figure obtained is a 3D histogram derived from the colours of the image. You
can rotate the figure around with the mouse to see the distribution of the colours.
One interesting observation is that the brown colours appear on both extremes of
the figure. This is due to the circular properties of the hue. Two other observations
are important. First, whilst on the original image it is easy to distinguish between the
brown and the blue colours, on the 3D distribution, however it is easy to observe that
there is a range from cyan to light purple. A rotation of the figure may help to see
this more clearly:

```
view(180, 2)
```

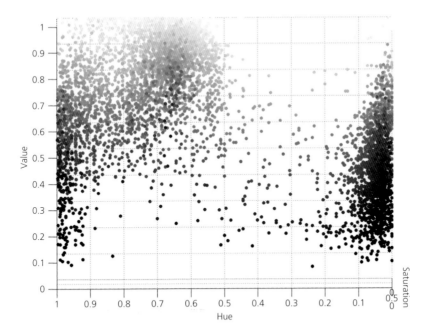

The second important observation is that there is a large number of pixels with very faint colours, near to white, that are hard to be perceived. A rotation and change of the background colour can highlight these elements:

```
set(gca, 'color', 'k')
view(38, 38)
```

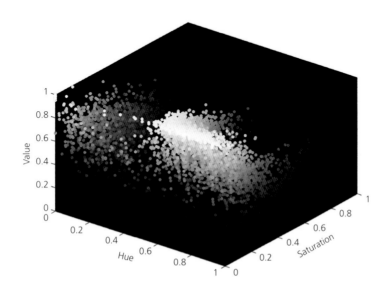

These 3D plots are very useful to explore the nature of colour images. We can introduce further properties to them to create multidimensional histograms, as will be explored in the following recipe.

6.7 Recipe 7: Multidimensional Histograms

One important aspect of quantitative analysis of images acquired through immuno-histochemistry is to quantify the cells that have been stained. One easy way is to measure the number of pixels that comply with certain parameters, for instance, the proportion of pixels with values greater than 0.5, as was done in the previous recipe. However, there are more techniques that can be used. In this recipe we will use histograms of different varieties to visualise distributions in more than one dimension.

A distribution that spans three dimensions is harder to visualise, so it is common to "project" the 3D distribution down to a 2D histogram. Likewise, it is possible to project a 2D histogram down to create a 1D histogram. In each of these projections, the dimension of interest is selected, and the others are implicit in the histograms. Technically, the lower dimension distributions are called "marginal distributions" of a higher dimensional distribution.

Marginal distributions of a 3D hue-saturation-value histogram correspond to 2D hue-saturation, hue-value or saturation-value histograms. Marginal distributions of 2D histograms correspond to 1D histograms: hue, saturation or value.

Let's begin with the simplest cases, the 1D histograms. These can be easily calculated from the separate channels calculated in the previous recipes with the function 'hist'. Notice that we will use the colon operator "(:)" after the matrix to rearrange the matrix into a single column, otherwise, 'hist' calculates the histograms for each individual column of the matrix.

```
[occurrencesHue, axisHue]     = hist(currentHue(:), 50);
[occurrencesSaturation, axisSaturation] = hist...
                         (currentSaturation(:), 50);
[occurrencesValue, axisValue]= hist(currentValue(:), 50);

relativeOccurrenceHue = occurrencesHue/sum(occurrencesHue);
relativeOccurrenceSat = occurrencesSaturation/...
                    sum(occurrencesSaturation);
relativeOccurrenceVal = occurrencesValue/...
                    sum(occurrencesValue);
```

The command 'hist' counts the number of elements that occur in 50 equally spaced slots or "bins" and stores it in the first output argument (i.e. 'occurrencesHue') and the centre of these bins in the second argument (i.e. 'axisHue'). We have divided the counts provided by the command 'hist' by the total number of pixels to obtain the ratios. Now we can display the occurrences in 1D:

```
figure
subplot(311)
```

```
plot(axisHue, relativeOccurrenceHue)
grid on
xlabel('Hue')
ylabel('Relative Occurrences')
subplot(312)
plot(axisSaturation, relativeOccurrenceSat)
grid on
xlabel('Saturation')
ylabel('Relative Occurrences')
subplot(313)
plot(axisValue, relativeOccurrenceVal)
grid on
xlabel('Value')
ylabel('Relative Occurrences')
```

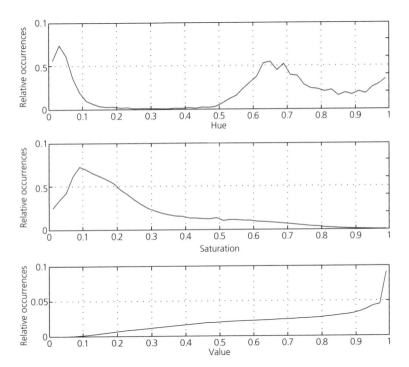

Now, let's observe the 2D distributions. We can approximate the 2D histograms from the 1D histograms that we just obtained by multiplying the vectors in the correct way (a column vector times a row vector). To make the figure more interesting, we will display again the range of hues and saturation underneath the 2D distribution.

```
figure
surf(xData, yData, zData1-1, image1_RGB, 'EdgeColor', 'none');
hold on
```

```
mesh(xData, yData, occurrencesSaturation'*...
    occurrencesHue/sum(occurrencesHue)/...
    sum(occurrencesSaturation),...
    'facecolor', 'none')
axis tight
xlabel('Hue')
ylabel('Saturation')
zlabel('Relative Occurrences')
rotate3d on
```

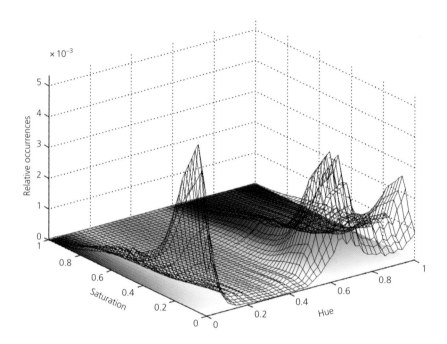

The distribution now confirms that the majority of the pixels of the figure have low saturations and are distributed between the cyan-blue region (0.5–0.8) and the red-purple region (0–0.1, 0.8–1). The combination with the values is the one that creates the brown colour, which is not evident in the current plane with the hues and saturation, as they have a value of 1.

Before we look at 3D distributions, it is useful to compare another image, for instance, one stained with Haematoxilin and Eosin (uploaded to Wikimedia Commons by 'Wars', 2 August 2006):

```
immunohisto_HandE     = imread(['http://upload.wikimedia.org',...
    '/wikipedia/commons/8/86/',...
    'Emphysema_H_and_E.jpg']);
immunohisto_HandE_hsv = rgb2hsv(immunohisto_HandE);
```

```
currentHue_HE           = immunohisto_HandE_hsv(:, :, 1);
currentSat_HE           = immunohisto_HandE_hsv(:, :, 2);
currentVal_HE           = immunohisto_HandE_hsv(:, :, 3);

[occurrencesHue_HE,axisHue]          = hist(currentHue_HE(:),50);
[occurrencesSaturation_HE,axisSaturation]   = ...
    hist(currentSat_HE(:),50);
[occurrencesValue_HE,axisValue]      = hist(currentVal_HE(:),50);

figure
imagesc(immunohisto_HandE)
figure
surf(xData, yData, zData1-1, image1_RGB, 'EdgeColor', 'none');
hold on
mesh(xData, yData, occurrencesSaturation_HE'*...
    occurrencesHue_HE/sum(occurrencesHue_HE)/...
    sum(occurrencesSaturation_HE),...'facecolor', 'none')
axis tight
xlabel('Hue')
ylabel('Saturation')
zlabel('Relative Occurrences')
rotate3d on
view(195, 25)
```

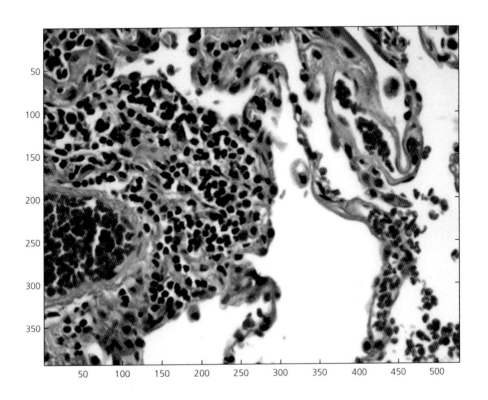

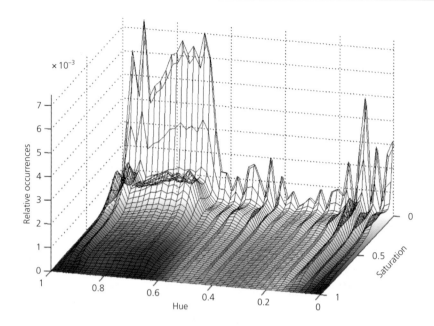

At first glance we can see that the distribution of the H&E image is also highly shifted towards the low saturation region. However, there is a large section of the image without any cells, so we could easily zoom in to remove that region, say 100:300 in the rows and 1:100 in the columns. In this way we can get a better representation of how H&E is distributed along hue and saturation. To do that we can repeat the plots but this time by selecting a subset of the image:

```
currentHue_HE              = immunohisto_HandE_hsv(100:300,...
                               1:200, 1);
currentSat_HE              = immunohisto_HandE_hsv(100:300,...
                               1:200, 2);
currentVal_HE              = immunohisto_HandE_hsv(100:300,...
                               1:200, 3);

[occurrencesHue_HE, axisHue] = hist(currentHue_HE(:), 50);
[occurrencesSaturation_HE, axisSaturation]  = hist(...
    currentSat_HE(:), 50);
[occurrencesValue_HE, axisValue]        = hist(...
    currentVal_HE(:), 50);

figure
imagesc(immunohisto_HandE(100:300, 1:200, :))
figure
surf(xData, yData, zData1-1, image1_RGB, 'EdgeColor', 'none');
hold on
mesh(xData, yData, occurrencesSaturation_HE'*...
    occurrencesHue_HE/sum(occurrencesHue_HE)/...
    sum(occurrencesSaturation_HE),...
    'facecolor', 'none')
```

```
axis tight
xlabel('Hue')
ylabel('Saturation')
zlabel('Relative Occurrences')
rotate3d on
view(195, 25)
```

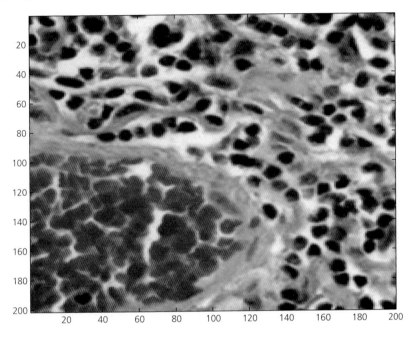

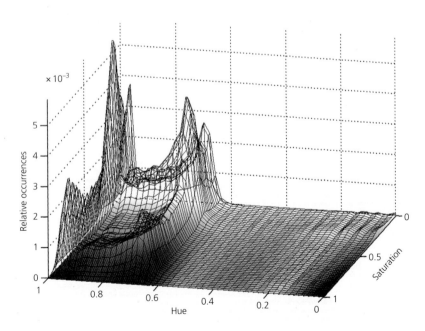

We now obtain a more representative histogram. If we compare this with the CD31 histogram, it is clear how the hues below 0.6 are completely absent. Furthermore, there are two large peaks, or rather ridges, at 0.7 and 0.96, in the hue dimension with a trough approximately at 0.92 and two small peaks around 0.82 and 0.86. Therefore we can use these values to separate into three regions of interest. We can also use the saturation to discard the bright regions that correspond to background.

First we need to zoom out and recalculate the matrices with the separate channels:

```
currentHue_HE          = immunohisto_HandE_hsv(:, :, 1);
currentSat_HE          = immunohisto_HandE_hsv(:, :, 2);
currentVal_HE          = immunohisto_HandE_hsv(:, :, 3);
```

And with these results, we can combine the saturation and the hues to distinguish the three regions of interest: dark purple cells, intense pink cells and the lighter pink cells that surround the other cells:

```
intensePink_cells      = (currentSat_HE(:, :, 1)>0.5).* ...
    (currentHue_HE(:, :, 1)>0.92);
lightPink_cells        = (currentSat_HE(:, :, 1)>0.15).* ...
    (currentHue_HE(:, :, 1)>0.81).* (currentHue_HE(:, :, 1)...
    <0.92);
purple_cells           = (currentSat_HE(:, :, 1)>0.5).* ...
    (currentHue_HE(:, :, 1)<0.81);
```

We can now display the output of the three segmentations, each in a separate axis.

```
figure
subplot(131)
imagesc(intensePink_cells)
subplot(132)
imagesc(purple_cells)
subplot(133)
imagesc(lightPink_cells)
colormap(gray)
```

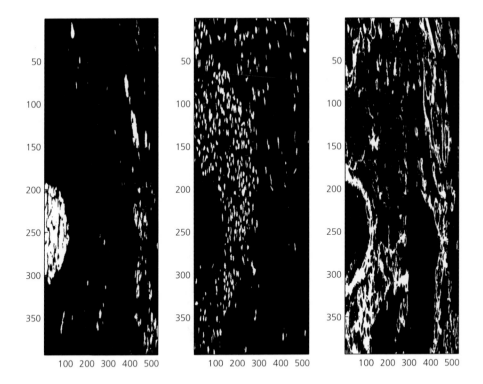

To combine the previous results with the original image, we need to perform a small trick. The original image is stored in 'uint8', which cannot be used for multiplications and other mathematical operations without loss of precision, so we will convert it into 'double'. Then, to normalise the values to the range 0 to 1, the pixel values are divided over the maximum value, in this case 255.

```
immunohisto_HandE_2      = double(immunohisto_HandE)/255;
```

Also, since the results are 2D matrices and the image is a 3D matrix, we need to repeat the matrices three times to have the same dimensions. Now we can see the segmentation results overlaid in the original image.

```
figure
subplot(131)
imagesc(immunohisto_HandE_2.*repmat(intensePink_cells,...
  [1 1 3]))
subplot(132)
imagesc(immunohisto_HandE_2.*repmat(purple_cells, [1 1 3]))
subplot(133)
imagesc(immunohisto_HandE_2.*repmat(lightPink_cells, [1 1 3]))
```

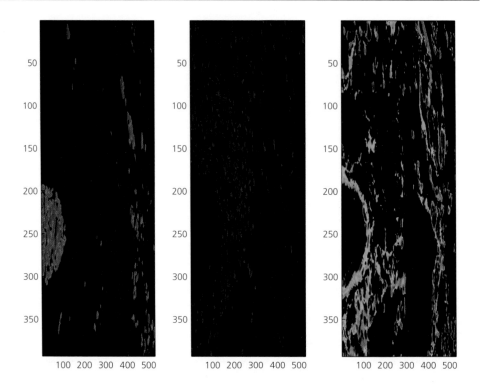

Three-dimensional histograms are not so easy to approximate from the 1D histograms, so we will calculate it "the long but simple way", that is by iterating over the possible values of each channel and counting how many fall into the combination of hue-saturation-values. To do this, it is necessary to have three loops, one for each channel. Then for each we will compare the channel against an upper and a lower edge, and then we will combine the current results and count the positions that fulfil the three ranges of values. Notice that these kinds of operations, where loops are used, take much longer than the internal Matlab functions.

```
edgesBins               = 0:0.1:1;
threeDhistogram(10, 10, 10)    = 0;
for counterHue=1:10
    disp(counterHue)
    huesBetweenEdges        = (currentHue>=edgesBins...
                              (counterHue)).* ...
                              (currentHue<edgesBins...
                              (counterHue+1));
    for counterSat =1:10
        satsBetweenEdges   = (currentSaturation>=edgesBins...
                              (counterSat)).*...
                              (currentSaturation<edgesBins...
                              (counterSat+1));
        for counterVal =1:10
            valsBetweenEdges = (currentValue>=edgesBins...
                              (counterVal)).*...
```

```
                                    (currentValue<edgesBins...
                                    (counterVal+1));
            combinedHSV         = huesBetweenEdges.*...
                                  satsBetweenEdges.*...
                                  valsBetweenEdges;
            threeDhistogram(counterHue, counterSat,...
                counterVal)= sum(combinedHSV(:));
        end
    end
end
```

```
    1

    2

    3

    4

    5

    6

    7

    8

    9

   10
```

Notice that we displayed the step of the counter with the command 'disp'. This is useful when we run long processes to have an idea of how long they will take to finish.

Finally, with the previously calculated 3D histogram, we can display in three dimensions by using the 'scatter3' function and assigning the size (fourth parameter) as proportional to the numbers in the 3D histogram. The colour will be calculated from the coordinates by transforming back from the HSV to the RGB space.

```
figure
hold on
for x=5:5:50
  for y=5:5:50
    for z=5:5:50
        scatter3(axisHue(x), axisSaturation(y), axisValue(z),...
            0.1+round(5e-3*threeDhistogram(x/5, y/5, z/5)),...
            hsv2rgb([x, y, z]/50), 'filled');
    end
```

```
   end
end
rotate3d on
view(143, 30)


xlabel('Hue', 'fontsize', 16)
ylabel('Saturation', 'fontsize', 16)
zlabel('Value', 'fontsize', 16)
```

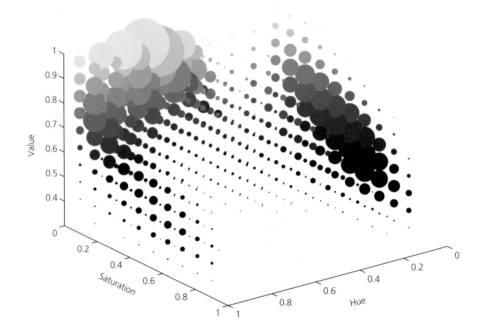

For display purposes, there is a small dot at every location of the histogram. This does not mean that there are pixels with those colours; it is simply used as a reference for the positions. We can compare the result now with the first 3D plot that we had obtained by subsampling the channels:

```
figure
scatter3(sampledHue, sampledSaturation, sampledValue, 20, ...
    [sampledRed, sampledGreen, sampledBlue]/255, 'filled');
xlabel('Hue', 'fontsize', 16)
ylabel('Saturation', 'fontsize', 16)
zlabel('Value', 'fontsize', 16)
view(143, 30)
```

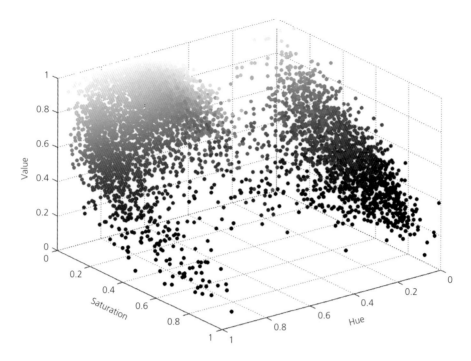

It should be clear now which colours are more or less present in the image, as clear faint spheres are much larger than any other colours.

Reference

Reyes-Aldasoro, C.C. 2009. "Retrospective Shading Correction Algorithm Based on Signal Envelope Estimation." Electronics Letters 45, no. 9: 454. doi:10.1049/el.2009.0320.

CHAPTER 7

Fluorescence Microscopy

Fluorescence, the emission of light by a substance that has absorbed light, is a very useful technique in the life sciences as it allows us to distinguish different biological molecules by associating them to different colours. Some molecules such as chlorophyll are naturally fluorescent, but it is possible to attach fluorescent dyes or "fluorophores" to other molecules like antibodies through a process called labelling. In this way, it is possible to label different parts of a cell with different fluorophores and the result is an image where different molecules can be identified by different colours.

In order to separate the signals emitted by different fluorophores, microscopes typically have a system of filters, commonly known as "cubes", "filter cubes" or "optical blocks". These cubes allow the light to "pass" through only for a certain range of colours, or more specifically for a certain range of frequencies of the electromagnetic spectrum. On the other hand, the filters will "block" other regions of the spectrum effectively obtaining separate images for each colour band. "Pass" and "block" should be understood in relative terms; "pass" will in fact be a small attenuation and "block" will be a higher attenuation of the light at a certain frequency.

Microscope manufacturers provide a large number of filter cubes with different optical characteristics that can target a certain frequency range or colour. Conversely, there is an even larger range of fluorophores available that will vary in their characteristics such as excitation and emission wavelengths, extinction coefficients or quantum yield.

As a first example, the following code will read and display a fluorescent image from Wikimedia Commons (file uploaded by "Benutzer:Jan R". 4 December 2005):

```
FluorescentCells_1 = imread(['http://upload.wikimedia.org/',...
   'wikipedia/commons/0/09/',...
   'FluorescentCells.jpg']);
imagesc(FluorescentCells_1)
```

Biomedical Image Analysis Recipes in MATLAB®: For Life Scientists and Engineers, First Edition.
Constantino Carlos Reyes-Aldasoro.
© 2015 John Wiley & Sons, Ltd. Published 2015 by John Wiley & Sons, Ltd.

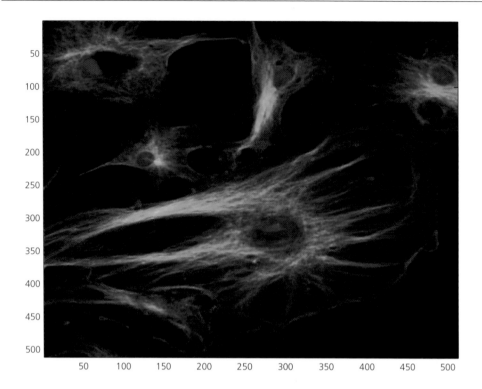

The image contains endothelial cells where the nuclei of the cells have been stained in blue, microtubules appear in green and the actin filaments are red.

The image is formed as a 3D matrix with 512 rows, 512 columns and 3 layers or levels.

```
size(FluorescentCells_1)
```

```
ans =

    512    512      3
```

Each element of the matrix will contain a number related to the intensity of the fluorescence, and this intensity is related to the number of photons detected by the acquisition device (that is the camera in the case of microscopy). The number of photons detected is in turn dependant on the abundance of the fluorophore in a certain region of the sample, and that is related to the presence of the molecules of interest.

The value of the intensity itself can be stored in many ways depending on how the number is recorded and depends on the range of values that the number can take (for instance 0 to 255 or 0 to 65536), if it can be positive or negative or if it contains decimal values. In the case of images, a common format or "data type" in Matlab is 'uint8', which translates to "8-bit unsigned integer", or in other words it can store values between 0 and 255 (that is the 8-bit part), no negatives (no signs or all positive) and no decimal points (integer). To check which data type is used in the image you can use the command 'class' in the following way:

```
class(FluorescentCells_1)
```

```
ans =

uint8
```

'uint8' is a common data format, but there are many others: 'uint16' is similar to 'uint8', but instead of 8-bit, it is 16-bit, which means it can represent numbers that go from 0 to 65536, and 'int8' is a signed integer of 8 bits, so the values can go from -128 to 127. In practice the data type restricts the range of values of a matrix and the operations you can do with those values. For instance, if your matrix is formed of 'uint8s', you can add, subtract, divide, etc. but the values will always be integers and between the range of 0 to 255. Look at the following example:

```
% Define 'a' with a value of 150 and force the type to 'uint8'
a = uint8(150);
% Define 'b' with a value of 120 and force the type to 'uint8'
b = uint8(120);
% 150 - 120 is 30, which is within the range of 'unit8'
a-b
% 120 - 150 is -30, not within the range of 'uint8'
b-a
% 150 + 120 is 270, not within the range of 'uint8'
a+b
% 150 * 120 is 18,000, not within the range of 'uint8'
a*b
% 150/120 is 1.25, not within the range of 'uint8'
a/b
```

```
ans =

    30
```

```
ans =

    0
```

```
ans =

    255
```

```
ans =

   255


ans =

   1
```

As shown in the previous example, 'uint8' is not a good data type to perform standard mathematical operations as it may give undesired results. The data type 'double', which is the one that Matlab uses by default when creating a variable, is more suitable as it can accommodate a very large range of numbers as well as decimals. The data type 'double' is a "double precision floating-point array" representation of value; floating-point means that number is an approximation of its real value and it stores a limited number of "significant digits", also called "mantissa", and an exponent to scale the significant digits. In other words, if a number is very large or very small, the exponent will keep track of the number of decimal places, however, a number of least significant decimal digits will be discarded and the mantissa will store the most significant ones. Look at the following example:

```
c=
 (1111111111111111111111111111111111.1111111111111111111111111111111)


  c =

    1.1111e+34
```

The variable 'c' is a 'double' and is stored as a mantissa with the most significant digits (in this case the long sequence of ones) and the exponent of 34.

In practice you do not need to worry too much about mantissas or exponents, just remember that if you are to do any kind of operations it is generally better to use 'double' rather than 'uint8'.

7.1 Recipe 1: Separating and Combining Colour Channels

The colours of the images are formed, in Matlab and other software packages, by combining colours in an "additive" way, which is how colours are formed in photography, televisions and computer monitors. The primary colours of the additive colour model are red, green and blue. Conversely, the primary colours of the subtractive colour model, which is commonly used in painting or mixing inks, are cyan, magenta and yellow. Therefore a 3D matrix with three levels will be interpreted as having red, green and blue colour components. This gives the possibility of creating colour images from individual grey-scale ones. As an example, first let us decompose a colour image into individual channels:

```
% select the first level of the 3D matrix
channel_1 = FluorescentCells_1(:,:,1);
```

```
% select the second level of the 3D matrix
channel_2 = FluorescentCells_1(:,:,2);
% select the third level of the 3D matrix
channel_3 = FluorescentCells_1(:,:,3);

figure
subplot(1,3,1); imagesc(channel_1)
subplot(1,3,2); imagesc(channel_2)
subplot(1,3,3); imagesc(channel_3)
colormap(gray)
```

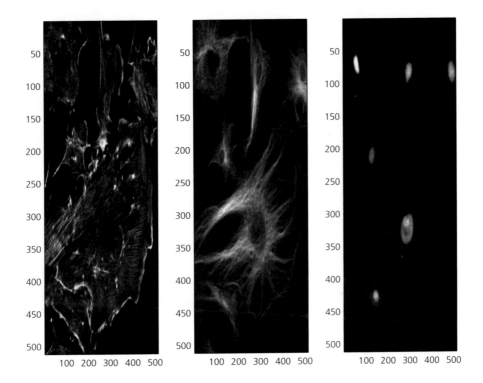

The previous commands have displayed each channel on a separate plot; this will be used later to perform operations. Now let's rearrange the channels into new images:

```
re_arranged_fluorescence_1(:,:,1) = channel_1;
re_arranged_fluorescence_1(:,:,2) = channel_3;
re_arranged_fluorescence_1(:,:,3) = channel_2;


re_arranged_fluorescence_2(:,:,1) = channel_3;
re_arranged_fluorescence_2(:,:,2) = channel_2;
re_arranged_fluorescence_2(:,:,3) = channel_1;
```

```
re_arranged_fluorescence_3(:,:,1)  =  channel_3;
re_arranged_fluorescence_3(:,:,2)  =  channel_1;
re_arranged_fluorescence_3(:,:,3)  =  channel_2;

figure
subplot(1,3,1);  imagesc(re_arranged_fluorescence_1)
subplot(1,3,2);  imagesc(re_arranged_fluorescence_2)
subplot(1,3,3);  imagesc(re_arranged_fluorescence_3)
```

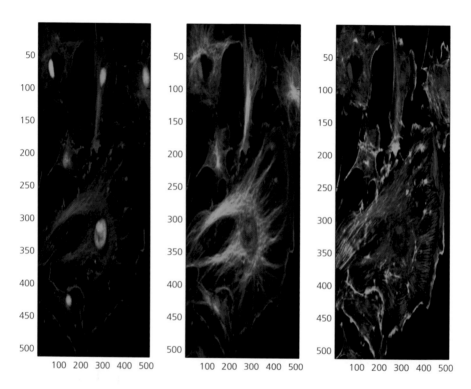

In the previous cases, the channels were swapped from their original place, and therefore what was blue became green, or what was red became blue. If we think of an individual location in rows and columns, or a single pixel, we can observe that the colour will correspond to the red, green and blue combination; this is sometimes called a "colour triplet" as it is formed by three values. To create colours other than red, green and blue, it is possible to modify the triplets by combining their values. For example what we see as green would correspond to a triplet with a high value in green and low values (ideally 0) in red and blue; if a triplet has high values in green and red and low blue it would look yellow, or high in green and blue would correspond to cyan and high in all channels would tend to white. These combinations are created in the following way with our example image:

```
re_arranged_fluorescence_4(:,:,1)  =  channel_1+channel_2;
re_arranged_fluorescence_4(:,:,2)  =  channel_2;
re_arranged_fluorescence_4(:,:,3)  =  channel_3;
```

```
re_arranged_fluorescence_5(:,:,1) = channel_1;
re_arranged_fluorescence_5(:,:,2) = channel_2;
re_arranged_fluorescence_5(:,:,3) = channel_3+channel_2;

re_arranged_fluorescence_6(:,:,1) = channel_1+channel_2;
re_arranged_fluorescence_6(:,:,2) = channel_2;
re_arranged_fluorescence_6(:,:,3) = channel_3+channel_2;


figure
subplot(1,3,1); imagesc(re_arranged_fluorescence_4)
subplot(1,3,2); imagesc(re_arranged_fluorescence_5)
subplot(1,3,3); imagesc(re_arranged_fluorescence_6)
```

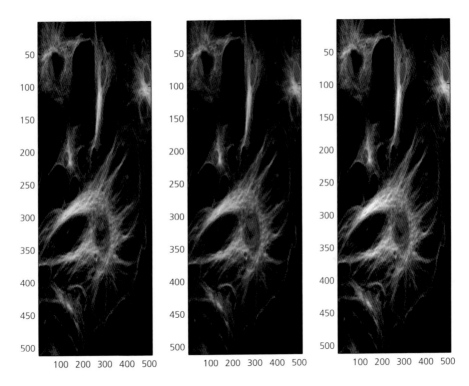

The previous image shows how the original green channel was modified to be displayed with different colours; of course, many other combinations are possible. A word of caution: as mentioned previously, the images are formed by 'uint8', which have a maximum value of 255. When the 'uint8's of the channels are added, they are capped at 255, or in other words, the image will become saturated at 255. Look for example at what happens when the image channels being multiplied by 4:

```
re_arranged_fluorescence_7(:,:,1) = 4*channel_1;
re_arranged_fluorescence_7(:,:,2) = 4*channel_2;
re_arranged_fluorescence_7(:,:,3) = 4*channel_3;
```

```
figure
imagesc(re_arranged_fluorescence_7)
```

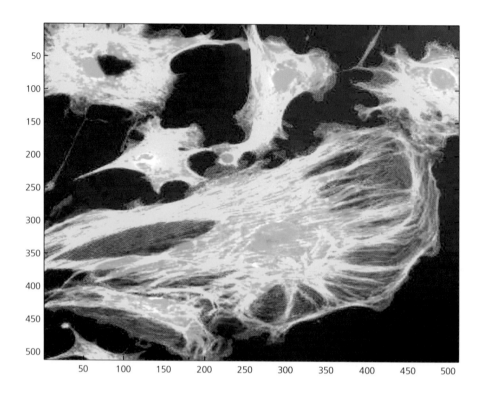

The colours have shifted towards white, and the detail of the image is lost. In future examples we will see how to avoid the saturation by using 'double'.

7.2   Recipe 2: Investigating the Scaling of Values

So far, we have used the command 'imagesc' to display images on a figure. Imagesc stands for "image-scale"; that is, before displaying the data, its intensity values are scaled. This provides advantages and disadvantages depending on what we are interested in doing. To exemplify this, let's look at another image (uploaded to Wikimedia Commons by "Masur" 29 June 2006):

```
FluorescentCells_2 = imread(['http://upload.wikimedia.org/',...
  'wikipedia/commons/a/ad/',...
  'Yeast_membrane_proteins.jpg']);
figure
imagesc(FluorescentCells_2)
```

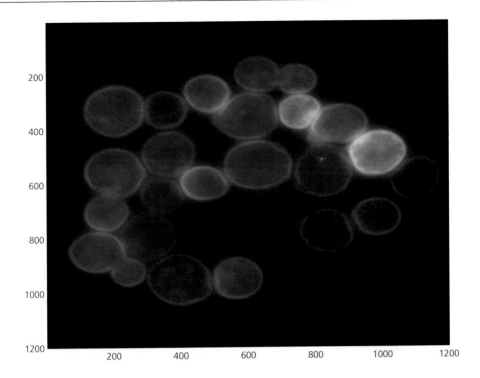

This image contains yeast cells whose membrane has been stained with two different fluorophores: Green Fluorescent Protein (GFP) and Red Fluorescent Protein (RFP). Different levels of expression of each protein create a range of colours from red to green with intermediate yellows. It should be clear by now that there is no blue component in this image. From an acquisition point of view there is no fluorophore associated to blue, and from a visual point of view, all colours are in the range of red, green and its combinations. To confirm this, we can again display the individual channels:

```
% select the first level of the 3D matrix
channel_1 = FluorescentCells_2(:,:,1);
% select the second level of the 3D matrix
channel_2 = FluorescentCells_2(:,:,2);
% select the third level of the 3D matrix
channel_3 = FluorescentCells_2(:,:,3);


figure
subplot(1,3,1); imagesc(channel_1)
subplot(1,3,2); imagesc(channel_2)
subplot(1,3,3); imagesc(channel_3)
colormap(gray)
```

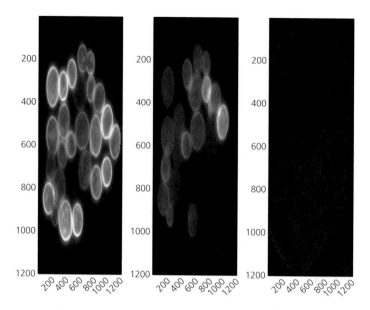

From the channels it is easy to detect that some cells are purely red and that they do not appear in the green channel, and the green cells have some red component. However, the blue channel shows a large number of pixels that appear to be of high intensity, which was not expected from what was explained above. This is due to the scaling effect, which is part of 'imagesc'. Let's add a scale bar (or colour key) to each plot so that we can see the range of values that each channel contains:

```
figure
subplot(1,3,1); imagesc(channel_1); colorbar
subplot(1,3,2); imagesc(channel_2); colorbar
subplot(1,3,3); imagesc(channel_3); colorbar
colormap(gray)
```

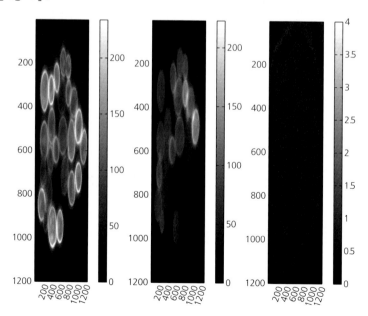

Whilst the red and green channels have intensities well above 200, the blue channel intensity is below 4. This is most likely noise, unwanted values that are intrinsically linked to the acquisition of the data. To control the scaling, it is possible to specify the limits in the range of values in which we are interested, in this case 0 and 255 as lower and upper limits, respectively. These limits are passed as a second argument to 'imagesc' in the following way:

```
figure
subplot(1,3,1); imagesc(channel_1,[0 255]); colorbar
subplot(1,3,2); imagesc(channel_2,[0 255]); colorbar
subplot(1,3,3); imagesc(channel_3,[0 255]); colorbar
colormap(gray)
```

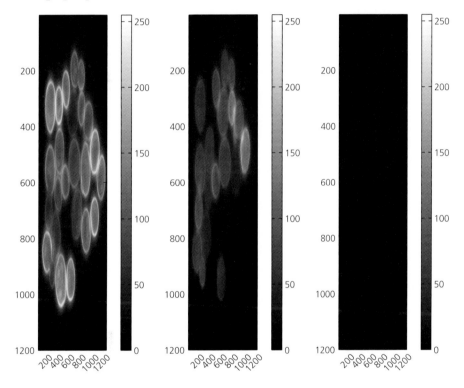

We can now see that the red and green channels have not changed considerably, as the intensities were high; however, the blue channel appears black, as should be expected from the low intensities.

7.3 Recipe 3: Automatic Threshold Selection

In previous sections we have used a segmentation technique called "thresholding" to separate the background from some objects of interest. This is a fairly simple and effective technique in which everything above a certain value is considered to be an object, with the slight complication of determining the actual value of the threshold. Whilst setting a threshold manually for one image can be done relatively fast, if there are hundreds of images to analyse, the task may become very time consuming and susceptible to intra- and inter-observer variability.

An automatic technique to determine a threshold that will separate an object or series of objects from the background was proposed in 1979 by Nobuyuki Otsu and it is widely used. The technique is based on statistical analysis of the images to maximise the separation between the object and the background. In Matlab, Otsu's algorithm is used through the function 'graythresh'. Using the previous example, let's threshold the red and green channels using 'graythresh':

```
FluorescentCells_1 = imread(['http://upload.wikimedia.org/',...
  'wikipedia/commons/0/09/',...
  'FluorescentCells.jpg']);

% select the first level of the 3D matrix
channel_1 = FluorescentCells_1(:,:,1);
% select the second level of the 3D matrix
channel_2 = FluorescentCells_1(:,:,2);
% select the second level of the 3D matrix
channel_3 = FluorescentCells_1(:,:,3);

level_1 = graythresh(channel_1)
level_2 = graythresh(channel_2)
level_3 = graythresh(channel_3)

        level_1 =

            0.2078


        level_2 =

            0.2745


        level_3 =

            0.2941
```

Notice that the results of 'graythresh' are numbers between 0 and 1 and not within the 'uint8' range of [0–255]. These threshold values have been normalised to be between the range of 0 to 1, that is, the results will always be between 0 and 1 regardless of the input values. We will later see that normalisation is a very powerful technique. When we are interested in further processing steps we might be interested in knowing the exact value in the [0–255] range in which we are thresholding the images. A common case is when we are not satisfied with the automatic threshold and we want to vary it slightly. In this case we need to de-normalise the threshold values by multiplying them by the value that was used to normalise them, in this case 255:

```
level_1 = 255*graythresh(channel_1)
level_2 = 255*graythresh(channel_2)
level_3 = 255*graythresh(channel_3)
```

```
level_1 =

    53


level_2 =

    70


level_3 =

    75
```

There are different ways to display the thresholded output. One of them is using a built-in function of Matlab called 'im2bw', which is short for image-to-black-and-white. However, once we have obtained the threshold values in the 'uint8' range, we can simply threshold the images by using the greater than operator (>).

```
figure
subplot(2,3,1); imagesc(channel_1);
subplot(2,3,2); imagesc(channel_2);
subplot(2,3,3); imagesc(channel_3);
subplot(2,3,4); imagesc(channel_1>level_1);
subplot(2,3,5); imagesc(channel_2>level_2);
subplot(2,3,6); imagesc(channel_3>level_3);
colormap(gray)
```

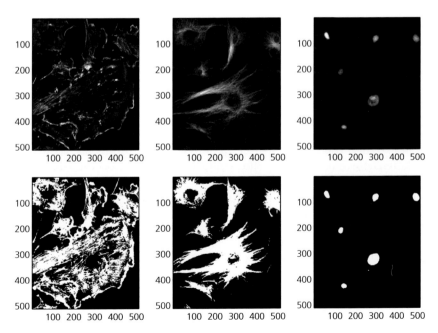

7.4   Recipe 4: Measuring Absolute and Relative Areas

In previous sections we introduced the concept of labelling, in which we assign a unique label to a region of connected pixels (i.e. those with values other than 0) of an image. Then we counted the number of elements and measured their respective areas. We will now explore further the measurements that we can extract from a labelled image using several other tools.

The first measurement that can be very simply derived from the thresholded images we just obtained is the relative area covered by the molecule of interest. As there are three channels, we can obtain three measurements of area, which are obtained by counting the number of pixels above the threshold. This count is equivalent to adding those pixels' numbers:

```
areaChannel_1        = sum(sum(channel_1>level_1))
areaChannel_2        = sum(sum(channel_2>level_2))
areaChannel_3        = sum(sum(channel_3>level_3))
```

```
         areaChannel_1 =

              102245


         areaChannel_2 =

               65649


         areaChannel_3 =

               5224
```

Notice that the function 'sum' was used twice for each channel. As each function will operate over one dimension, the first sum will add the pixels of each column, leaving a single row. The second sum will add over the row, leaving a single value. When the data has three dimensions three sums are required to add overall dimensions. Alternatively, the colon operator can be used to rearrange a matrix of any dimensions into a 1D matrix; in that case a single function sum is necessary:

```
areaChannel_1                 = sum(channel_1(:)>level_1)
```

```
      areaChannel_1 =

              102245
```

The absolute number of pixels is always relative to the size of the image, so it is a better idea to express it in relative terms. To do this, we can divide the sum of the pixels by the dimensions of the image:

```
[rows,columns,levels]    = size(FluorescentCells_1);
```

```
relativeAreaChannel_1    = areaChannel_1/rows/columns
relativeAreaChannel_2    = areaChannel_2/rows/columns
relativeAreaChannel_3    = areaChannel_3/rows/columns
```

```
    relativeAreaChannel_1 =

        0.3900


    relativeAreaChannel_2 =

        0.2504


    relativeAreaChannel_3 =

        0.0199
```

In this example, the number of pixels of the red channel that have an intensity higher than the Otsu level are approximately 39%, the green channel 25% and the blue channel 2%. These measurements are useful when we want to compare two cell populations, which may be expressed differently, and this can be estimated by the proportional areas they cover in the image.

7.5 Recipe 5: Counting Nuclei

The basic labelling procedure was introduced previously. It will be presented here as a starting point for more sophisticated measurements.

To count the number of nuclei, and later extract measurements from them, we need to assign a unique label to each of them. The command to label is 'bwlabel', and to obtain basic properties the command is 'regionprops':

```
[Nuclei,numNuclei]= bwlabel(channel_3>level_3);
areaNuclei          = regionprops(Nuclei,'Area');
figureTitle ...
  = strcat('Number of Nuclei:' ,num2str(numNuclei),...
         ', areas = [',num2str([areaNuclei.Area]),']');
figure
imagesc(Nuclei)
title(figureTitle)
```

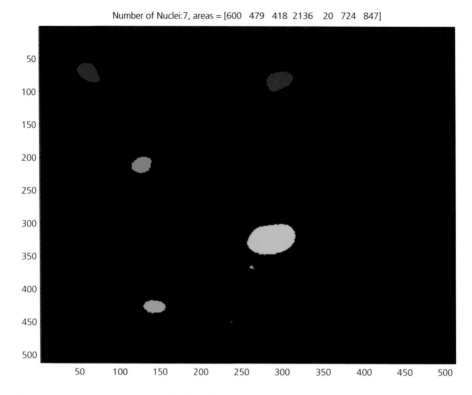

Number of Nuclei:7, areas = [600 479 418 2136   20 724 847]

The previous commands labelled the segmented nuclei, obtained the area of each nuclei, displayed the labelled image with the number of nuclei and the respective areas as the title of the figure. The areas are stored in the variable 'areaNuclei', which is of a data type called structure or 'struct' for short. Structures are a special type of variables or arrays, which contain multiple values. However, a 'struct' is different from a matrix in which you store multiple values in rows and columns because the values of the structure are identified by a unique name or "field". So, the variable 'areaNuclei' contains seven values of area stored in a single field called 'Area':

areaNuclei

        areaNuclei =

        7x1 struct array with fields:

                Area

If we want to observe a particular area value, we must specify it, for example:

areaNuclei(1)

```
    ans =

        Area: 600
```

And if we were to store that value of area in a different value, we would need to address the field directly using the dot (.) syntax like this:

```
b = areaNuclei(1).Area
```

```
    b =

        600
```

The use of square brackets around a struct, is a shortcut to extract all the values of a particular field:

```
[areaNuclei.Area]
```

```
    ans =

      Columns 1 through 6

          600       479       418      2136        20       724

      Column 7

          847
```

From the areas of the nuclei, we can observe that there is one object that is very small compared with the rest. This is located very close to the largest nuclei and we can consider that this object is not really a nucleus but an outlier, which should be removed before we proceed further. To remove the object, we will use the structure with the values for the areas and relabel only those objects with area larger than 30, as the object has an area of 20.

```
NucleiLarge    = bwlabel(ismember(Nuclei,...
   find([areaNuclei.Area]>30)));
figure
imagesc(NucleiLarge)
```

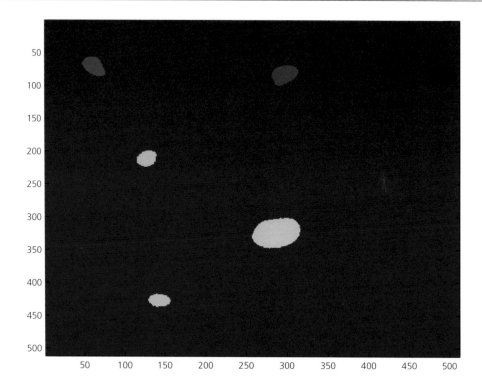

In an automated process we could use the mean and the standard deviation of the values of the area to select the outlier area threshold.

7.6   Recipe 6: Quantification of Region Properties Beyond the Area

Besides the area, there are many other useful measurements that can be obtained with the function 'regionprops', for instance:
• perimeter around an object, which is useful to obtain the area / perimeter ratio
• eccentricity, which measures how round or elongated an object is
• centroid, or the coordinates of the centre of the object

```
propertiesNuclei_1    = regionprops(NucleiLarge,'Area',...
                        'Centroid','Eccentricity');

propertiesNuclei_1(1)


    ans =

            Area:  600
        Centroid:  [61.2800 70.9250]
    Eccentricity:  0.7539
```

If we pass as an input argument the original image together with the labelled image we can also obtain some interesting measurements related to the regions and the intensity of the pixels, namely the maximum, minimum or average intensity of the object.

```
propertiesNuclei_2 = regionprops(NucleiLarge,channel_3,...
                     'MaxIntensity','MinIntensity',...
                     'MeanIntensity');

propertiesNuclei_2(1)


        ans =

            MeanIntensity: 201.1633
             MinIntensity: 76
             MaxIntensity: 255
```

A good way to analyse the measurements of our objects is to display them in a graphic way. In the following lines we use the command 'plot3' to display a red circle for each nucleus. The circles will be placed according to their characteristic values of (1) area, (2) eccentricity and (3) intensity.

```
figure
plot3([propertiesNuclei_1.Area],...
    [propertiesNuclei_1.Eccentricity],...
    [propertiesNuclei_2.MeanIntensity],'ro','markersize',9)
grid on
axis tight
rotate3d on
xlabel('Area');
zlabel('Intensity')
ylabel('Eccentricity')
```

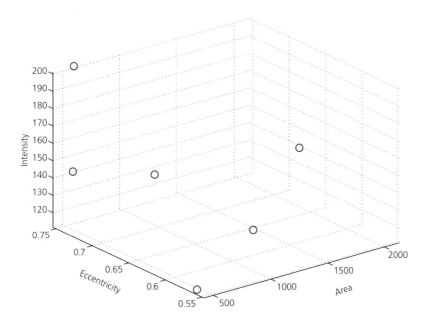

This kind of plot is useful to observe characteristics that may or may not be evident from the original image. For instance, we can see that there is one nuclei much bigger than the others (right side of the plot) and one nuclei brighter than the others (upper region of the plot). Whilst these observations may be evident from the image, the roundness or eccentricity may be harder to analyse from the images but are displayed clearly visible in the plot where two objects are rounder (closer to 0) than the other four objects.

7.7   Recipe 7: Dividing an Image into Regions

One important processing step in image analysis is the possibility of dividing an image into regions given a certain criterion, region uniformity for example. In the case of fluorescent cells, it may be interesting to divide the image into regions that contain a single cell. Following the previous recipes, it may be tempting to divide the image based on the nuclei that have been identified.

A popular mathematical technique to divide a space into regions is the "Voronoi tessellation" or "Voronoi diagram". The technique is based on a series of points or seeds, and the space is divided so that every point in the image is assigned to the closest seed.

In the case of an image with cells, the centres of the nuclei can be used as the seed points. These centres can be easily calculated as the centroids with 'regionprops':

```
centroidsNuclei          = regionprops(NucleiLarge,'Centroid');
```

The variable 'centroidsNuclei' is a 'struct' with one field, which contains two values for the coordinates of the centroid of each nucleus, for instance:

```
centroidsNuclei(1)
```

```
    ans =

        Centroid: [61.2800 70.9250]
```

If we list all the elements of the 'struct' by converting the 'struct' to a matrix by placing it inside square brackets, the list will place all the coordinates in a single 1D matrix, one centroid after another:

```
centroidsNuclei_matrix   = [centroidsNuclei.Centroid]
```

```
    centroidsNuclei_matrix =

    Columns 1 through 7

      61.2800   70.9250   127.1670   210.9395   142.6005   426.6292
      286.4817

    Columns 8 through 12

      323.7837   295.6561   82.5401   491.9941   82.5502
```

Therefore, to separate these into rows and columns, we need to separate the odd and even elements of the matrix in the following way:

```
rows_Centroids           = centroidsNuclei_matrix(2:2:end);
columns_Centroids        = centroidsNuclei_matrix(1:2:end);
```

With the centroids in place as two separate matrices, we can now obtain the Voronoi tessellation with the command 'voronoi' in the following way:

```
figure
imagesc(channel_3)
hold on
voronoi(columns_Centroids,rows_Centroids)
colormap(gray)
```

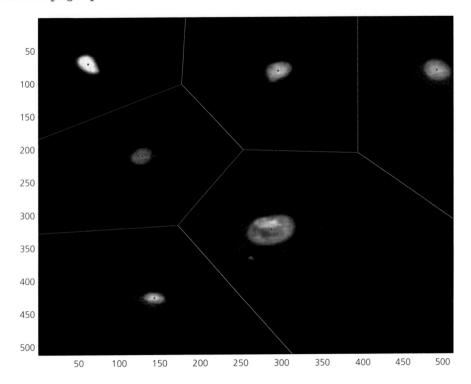

Notice that after displaying the image with the nuclei, we used the command 'hold on', in order to keep that image together with the Voronoi tessellation. These regions are based on the centroids of the nuclei and have disregarded the information contained in the other channels. This is manifest if we plot the tessellation over the channels themselves:

```
figure
imagesc(channel_2)
hold on
voronoi(columns_Centroids,rows_Centroids)
colormap(gray)
```

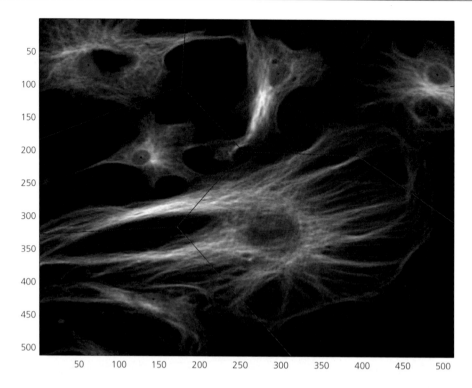

We can now see that the regions that have been obtained with the Voronoi tessellation do not fit accurately the distribution of the cells in the image. If we begin with the green channel we may get better results. First we should segment the channel into objects and background with the automatically calculated threshold and label the resulting objects:

```
channel_2                     = FluorescentCells_1(:,:,2);
level_2                       = 255*graythresh(channel_2);
[LabActin, numActinRegions]   = bwlabel(channel_2>level_2);
```

In this case, the number of regions is rather large, 199 for this particular image, as there are many small actin filaments that may be segmented into many disjointed objects. However, we know that there are six cells and we can take advantage of this information. First, we calculate the area in the same way as before.

```
areaActin       = regionprops(LabActin,'Area');
```

Then, we will select the six objects with the largest areas, as each of these must correspond to one cell. To select the largest areas we need to sort them, identify which of the labelled regions correspond to them and then discard the rest of the regions.

```
sortedArea      = sort([areaActin.Area]);
largestAreas    = find([areaActin.Area]>sortedArea(end-6));
LabActinLarge   = (ismember(LabActin,largestAreas));
```

```
figure
imagesc(LabActinLarge)
colormap(gray)
```

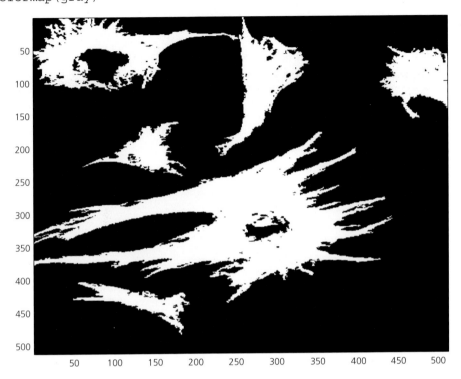

The function 'sort' orders from smallest to largest all the elements in a matrix. We can then select the seventh largest value and threshold the same matrix so that the positions with the six largest areas are greater than the threshold. The function 'find' then finds those values and returns their indices so that we can keep only those objects from the image. The function 'ismember' is the one that discards certain objects from a labelled image.

Once we have obtained six separate regions from the green channel, we can use a very useful segmentation technique called the "watershed transformation". In a topographical analogy with the distance corresponding to the altitude of a map, the object constitutes a "plateau" while the background was a descending slope. If rain falls on this artificial landscape, the water will find the path through which it could reach a region of minimum altitude, sometimes called a lake or sea. Each point in the map corresponded to a path towards one and only one lake, and the landscape was partitioned into catchment basins or regions of influence of the regional minima. Finally, the boundaries defined by the watershed can be used to partition the one object into a series of smaller objects. In our example, we want to partition the background, and the actin regions will be the catchment basins. We can use the command 'watershed' in the following way:

```
segmentedBackground      = watershed(1-LabActinLarge);
```

```
figure
imagesc(segmentedBackground)
```

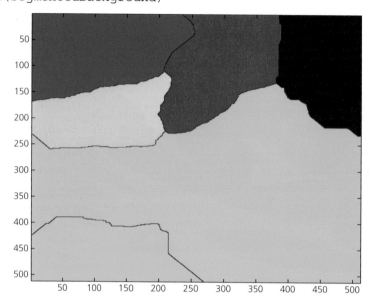

Notice that we have inverted the segmentation before we obtained the watersheds by subtracting the image from the value 1. The effect of this intensity inversion can be observed if we obtain the watershed without that subtraction:

```
segmentedBackground2      = watershed(LabActinLarge);
figure
imagesc(segmentedBackground2)
```

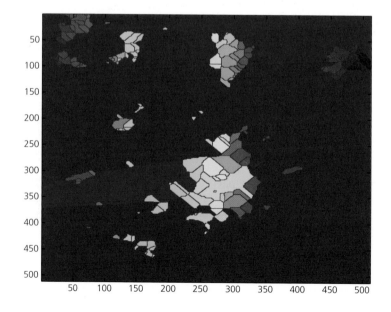

It should be clear now that the watershed obtained with the subtraction follows more accurately the distribution of the cells as the watershed without the inversion is dividing the regions internally, which is not what we are trying to do.

The boundaries defined by the watershed transformation are the ones that have partitioned the region. For purely visual purposes we can dilate these boundaries so that they appear wider on the results. Since we will later combine the boundaries with the channels that are 'uint8', it is practical to convert the boundaries to 'uint8' as well:

```
segmentedBackground_dil = 255*uint8(imdilate...
                          (segmentedBackground==0,ones(3)));
```

Finally, to display the boundaries overlaid on the channels we can add the boundaries to each channel. For this we will need to convert the boundaries to 'uint8' to have the same data types.

```
channel1_segmented      = channel_1+segmentedBackground_dil;
channel2_segmented      = channel_2+segmentedBackground_dil;
channel3_segmented      = channel_3+segmentedBackground_dil;

figure
subplot(131)
imagesc(channel1_segmented);
subplot(132)
imagesc(channel2_segmented);
subplot(133)
imagesc(channel3_segmented);
colormap(gray)
```

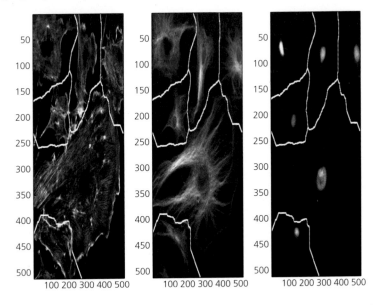

And those channels can be used to re-create the segmentation of the original image:

```
FluorescentCells_1_Segmented(:,:,1) = channel1_segmented;
FluorescentCells_1_Segmented(:,:,2) = channel2_segmented;
FluorescentCells_1_Segmented(:,:,3) = channel3_segmented;

figure
imagesc(FluorescentCells_1_Segmented)
```

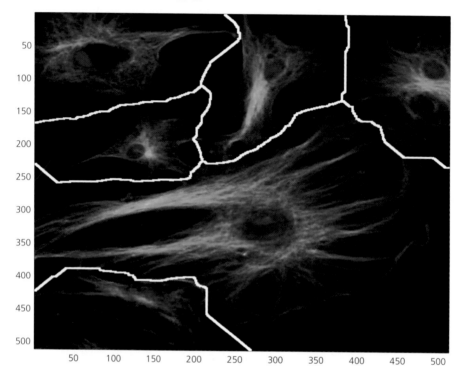

The improvement of the segmentation of the cells with this technique over the Voronoi tessellation is evident.

A final word of caution is necessary; as can be seen from the last image, the segmentation is not perfect as in some regions the segmentation lines are slightly too far from the real boundaries of the cells. It is therefore necessary to keep in mind that any numerical results associated with automated processes like the ones described in this recipe are an approximation of the true result. The actual advantage of algorithms comes when many images need to be processed. In those cases, algorithms can provide a consistent processing, free from intra- and inter-observer variability. The way to process many images in a single script is explained in the next recipe.

7.8   Recipe 8: Batch Processing and Montages

In many cases, we may be interested in processing not just one image, but a large number of images. Imagine for example the situation in which an experiment has been performed with several different parameters like cell lines, concentration of a drug or medium, temperatures, times of observation, etc. Experiments like this one may easily generate several hundreds of images, and it is in those cases where the power of automated or semi-automated algorithms can be fully exploited. The execution of a series of repetitive algorithms over a series of images is sometimes called "batch processing".

Matlab allows the processing of multiple images by processing them in "loops". Each image can be read, then it can be processed and the results saved into a new image, or measurements can be extracted and saved into a variable that can be used later for statistical analysis.

To illustrate the way to process multiple images, we will work with 14 synthetic images placed in a single folder:

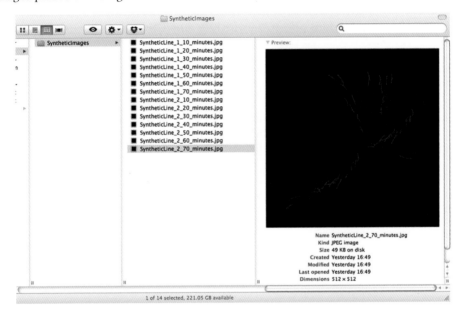

The first step is to read the images that are stored inside the folder and save the information in a variable that will be used later. The command 'dir' lists the files and folders in a given folder.

```
folderContents  = dir ('Datasets/SyntheticImages')

       folderContents =

       16x1 struct array with fields:

            name
            date
            bytes
            isdir
            datenum

```

It is important to highlight several issues at this stage. The first is that different operating systems interpret special symbols like %, &, * and even spaces in different ways. Although it is convenient for humans to use descriptive names with spaces and symbols, these may create problems when reading files in Matlab. Therefore, it is better to restrict file names to the uppercase and lowercase letters a to z, numbers and if necessary use underscore "_" to separate words, for instance 'Synthetic_image_1.jpg' is a good name, while '%this is & my \second image.bmp' is not. Second, sometimes it is convenient to use file names with some information about the images, like the time or the cell line. In these cases, it is very important to be consistent in the way these details are written as this information may be extracted from the file name, i.e. we should not use minutes in one case and hours in a different case. In the example, the name includes the number of the cell line, i.e. 1/2, and the minutes at which the images were acquired. Third, computers sometimes save "hidden files"; these may include small files related to the way folders are displayed. Matlab will read all hidden files; in the example, there are 14 images in the folder, and the variable 'folderContents' is a 'struct' with 16 records, that is, 14 for the images and 2 for hidden files. To avoid any problems with hidden files, it is better to list the folder contents specifying the kind of images to be read. This is done with a "wildcard". A "wildcard" or "wildcard character" is a symbol, normally the asterisk "*", which substitutes or matches one or more characters. For example, if we are interested in all the files that end with the letters "jpg", the following code '*.jpg' would be translated as "any file with extension jpg":

```
folderContents        = dir ('Datasets/SyntheticImages/*.jpg')


    folderContents =

    14×1 struct array with fields:

        name
        date
        bytes
        isdir
        datenum
```

The folder now contains 14 records, each one contains several fields, among which the name of the file is the one that will be used later on, for example:

```
folderContents(1)


    ans =

            name: 'SyntheticLine_1_10_minutes.jpg'
            date: '07-Nov-2013 12:20:00'
           bytes: 15013
           isdir: 0
         datenum: 7.3555e+05
```

Since we will need the number of images later on, we should store this value on a variable. This is obtained by calculating the size of the variable 'folderContents'.

```
numberOfImages          = size(folderContents,1);
```

If we are to read the images one by one we will need to concatenate the name of the folder where they are stored and the actual name of the file. The command 'strcat' concatenates two or more strings into a single one that can be later used, for example:

```
currentFile             = strcat('Datasets/SyntheticImages/',...
                          folderContents(1).name);
dataIn                  = imread(currentFile);
imagesc(dataIn)
```

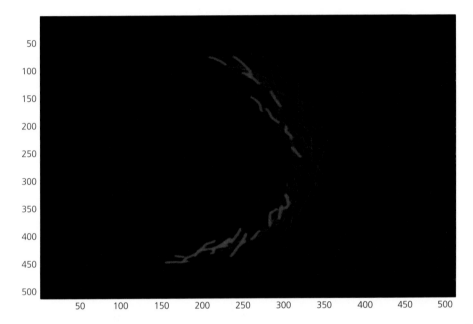

These instructions can be nested inside a loop that can read all the images on the folder. For simplicity, let's can assume that we are interested in counting the number of pixels on the red channel that are above the intensity level of 50. Thus we can process all the images on the folder with the following code:

```
for counterImages=1:numberOfImages
    currentImage        = folderContents(counterImages).name;
    currentFile         = strcat('Datasets/SyntheticImages/',...
                          currentImage);
    dataIn              = imread(currentFile);
    dataSegmented       = dataIn(:,:,1)>60;
    currentPixels(counterImages)= sum(sum(dataSegmented));
end
```

The 'for-loop' will iterate over the instructions inside the loop changing the value of the variable 'counterImages'. In this case, it will start with 1 and will increment by 1 until it reaches the value of 'numberOfImages', in this case 14. The instructions that follow will concatenate the names of the current image, read the image and store it in 'dataIn', then compare every pixel against a threshold of 60 and finally count the number of pixels above the threshold and store in the variable 'currentPixels'. Notice first that the counting is done using the command 'sum', which will act first on every column of 'dataSegmented', adding all the elements on each column, then the second sum will add all the values resulting in the total number of pixels above 60. Also notice 'currentPixels' is a 1D matrix, and the results of each image are stored in different positions. Then we can examine these results by looking at the values or plotting them like this:

```
figure
plot(currentPixels,'b-o')
```

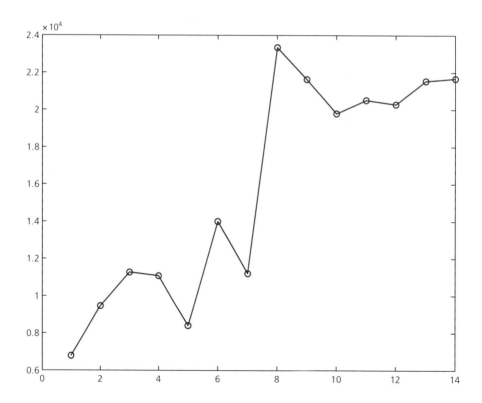

Let's now suppose that we want to analyse the results according to the two different groups; images (1–7) are group 1 and (8–14) are group 2. We can create a box plot with this command:

```
boxplot(currentPixels,[1 1 1 1 1 1 1 2 2 2 2 2 2 2]);
```

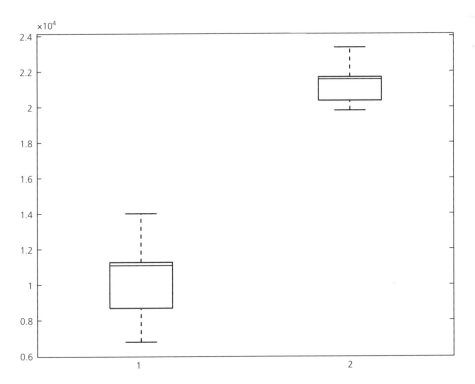

The groups have been assigned manually with ones and twos in a matrix. In the same way, we can test the statistical difference between the results with a t-test. The command 'ttest' is a one-sample and paired-sample t-test, and the command 'ttest2' performs a two-sample t-test, to check the independence of two samples. In this case, it is necessary to split the samples that will be passed as input to the command:

```
[rejectNullHypothesis,p_Value] =...
     ttest2(currentPixels([1:7]),currentPixels([8:14]))


     rejectNullHypothesis =

         1


     p_Value =

         1.2068e-07

```

For this simple example, it is easy to separate the groups for the 'boxplots' or the t-test by manually assigning them to a group. However, when the number of cases grows, it can be complicated to do this, and here is where a good naming system can be very useful. In the example the names of the images contain the cell line (1,2) and the time points (10–70 minutes). We can extract those values from the names as the for-loop reads each name. We then need to assign them to a new variable. The variable 'currentImage' stores the name of the image, and in the example, the 15th character corresponds to the cell line and the 17th–18th correspond to the time:

```
currentImage
currentImage(15)
currentImage(17:18)
```

```
currentImage =

SyntheticLine_2_70_minutes.jpg


ans =

2


ans =

70
```

Thus, we can run again the for-loop and store the corresponding values at each iteration. Furthermore, let's investigate not just one but two thresholds. Notice that the variable 'currentImage' is a string of characters and therefore to save as numbers these have to be converted with the command 'str2num' or 'str2double':

```
pixelThreshold                          = 60;
pixelThreshold2                         = 90;

for counterImages=1:numberOfImages
    currentImage   =...
       folderContents(counterImages).name;
    currentFile    =...
       strcat('Datasets/SyntheticImages/',currentImage);
    dataIn         = imread(currentFile);
    dataSegmented  = dataIn(:,:,1)>pixelThreshold;
    dataSegmented2 = dataIn(:,:,1)>pixelThreshold2;
    currentResults(counterImages,1) =...
          str2double(currentImage(15));
    currentResults(counterImages,2) =...
          str2double(currentImage(17:18));
    currentResults(counterImages,3) = sum(sum(dataSegmented));
    currentResults(counterImages,4) = sum(sum(dataSegmented2));
end
```

Now that the group and the times are stored in the first and second columns of 'currentResults' we can use these in the commands:

```
figure
boxplot(currentResults(:,3),currentResults(:,1));
title('Boxplots grouped by cell line')
```

```
figure
boxplot(currentResults(:,3),currentResults(:,2));
title('Boxplots grouped by time points')
```

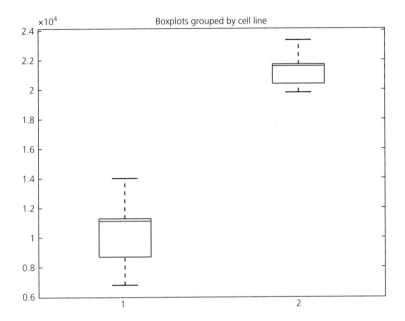

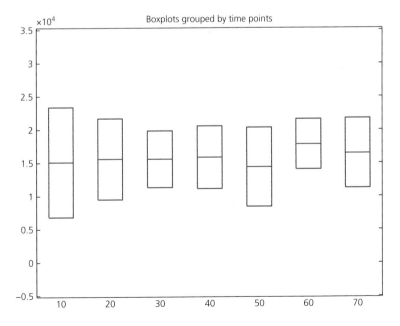

We can see that the measurements are statistically different when grouped by cell lines, however, they are fairly uniform when grouped by time points.

It is important to mention that statistical measures and plots should always be used with a significant number of samples. Here we display a boxplot with two samples as an illustration, but in real cases, more samples are necessary. When we have a low number of measurements, it is better to plot each cell line like this:

```
figure
hold on
plot (currentResults(currentResults(:,1)==1,2),...
    currentResults(currentResults(:,1)==1,3),'r-o')
plot (currentResults(currentResults(:,1)==1,2),...
    currentResults(currentResults(:,1)==1,4),'r--o')

plot (currentResults(currentResults(:,1)==2,2),...
    currentResults(currentResults(:,1)==2,3),'b-*')
plot (currentResults(currentResults(:,1)==2,2),...
    currentResults(currentResults(:,1)==2,4),'b--*')

grid on
xlabel('Time [minutes]');
ylabel('Num. Pixels with intensity > threshold');
legend(strcat('Cell Line 1, threshold =',...
  num2str(pixelThreshold)),...
    strcat('Cell Line 1, threshold =',...
    num2str(pixelThreshold2)),...
    strcat('Cell Line 2, threshold =',...
    num2str(pixelThreshold)),...
    strcat('Cell Line 2, threshold =',...
    num2str(pixelThreshold2)),...
    'location','north');
```

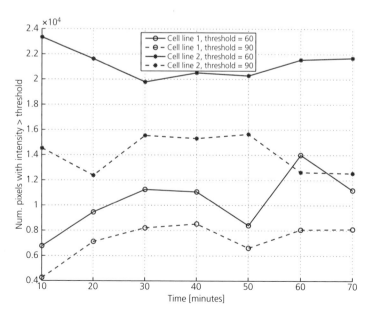

This final plot is very useful to observe the behaviour of the experiments against time, cell line and threshold level at the same time.

To create the plots, we have selected a subset of the matrix 'currentResults', which contains all the results. This is done by addressing the matrix with the first column that contains the group, i.e. 'currentResults(:,1)==1' corresponds to the first cell line and 'currentResults(:,1)==2' corresponds to the second one. We have also used the command 'legend', which adds a line of text that corresponds to each plot of the figure. These lines of text have to follow the same order as the plots. Another addition is the use of the three dots "...", which indicate to Matlab that the line is to be continued on the following line. This is very useful when writing long commands.

In many cases, once we display some characteristics of the images, it is important to verify that we are obtaining the correct measurements and the easiest way to do this is to visualise the input data, if possible all at once.

One convenient way of doing this is presenting several images as a montage, that is, arranging the images in a sequential order in a rectangular array. The images can be arranged in different ways; one of them is to have them as a multidimensional matrix, for example:

```
for counterImages=1:numberOfImages
    currentImage      = folderContents(counterImages).name;
    currentFile       = strcat('Datasets/SyntheticImages/',...
                        currentImage);
    dataIn2(:,:,:,counterImages) = imread(currentFile);

end
```

In the previous lines, we created a 4D matrix, with the 3D images placed one after another in the fourth dimension of 'dataIn2'.

With the images arranged into a single matrix, we can display them with the command 'montage'. Since we are interested in displaying each group of images separately, we can address the matrix in the fourth dimension so that we can plot each group in a different figure:

```
figure
montage(dataIn2(:,:,:,1:7))
title('Cell Line 1')
```

```
figure
montage(dataIn2(:,:,:,8:14))
title('Cell Line 2')
```

Cell line 1

Cell line 2

It is now clear that the results previously obtained are consistent with the visual analysis that we can perform with the montages; the images of the Cell Line 2 have more pixels over a certain threshold, than those of Cell Line 1.

7.9   Recipe 9: A Myriad of Measurements

As we have seen in Recipe 8, it is possible to read several images and process them with a series of steps, and then obtain measurements from the images and store them to perform statistical analysis with them. In some cases, where the images contain elements that can be individually analysed, like cells, we can further obtain measurements from each of these. This allows the possibility of generating a very large amount of measurements, which can reveal much about the nature of the experiments behind the images.

For this recipe we will analyse a series of synthetic images, which contain cells that have been stained in green. The cells, which we can assume are staining the walls of blood vessels, have different characteristics, which we will extract with different algorithms.

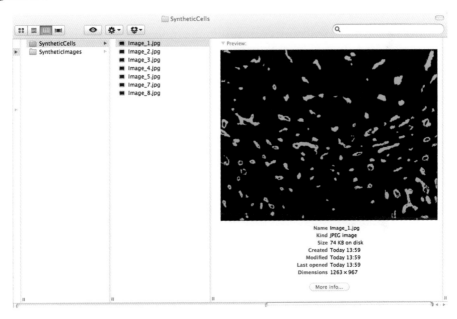

We can read the folder contents, obtain the number of images and read the first image as in the previous recipe:

```
folderContents = dir ('Datasets/SyntheticCells/*.jpg');
numberOfImages = size(folderContents,1);
currentFile    = strcat('Datasets/SyntheticCells/',...
                        folderContents(2).name);
dataIn         = imread(currentFile);
figure
imagesc(dataIn)
```

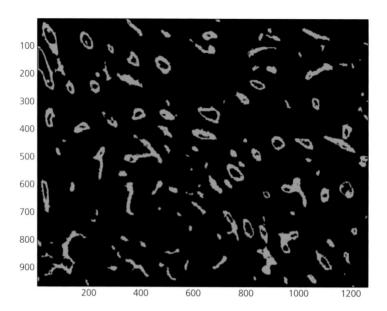

The first step to obtain measurements from the individual vessels is to distinguish them by assigning a unique label to each of them. This is done with the command 'bwlabel', and at the same time we can calculate the number of vessels in the image. In this case, we are only interested in the bright pixels of the green channel, therefore we can threshold and label them like this:

```
data_channel2                    = dataIn(:,:,2)>50;
[dataIn_Labelled,numberCells]    = bwlabel(data_channel2);
imagesc(dataIn_Labelled)
```

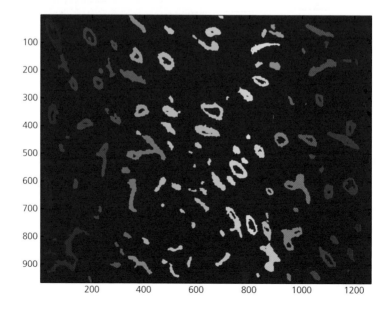

The vessels have now been assigned labels. In this example they span from 1 to 147. In the image, the labels correspond to different colours of the 'jet' colormap. In some cases, these colours are useful, but in others, it may be useful to have vessels that are close to each other with different colours. We can change the way the image is displayed by manipulating the colour maps. First let's change the background so that it is black instead of blue. We will copy the colour map jet into a new variable and then alter the first row, which corresponds to the background:

```
jet2                   = jet;
jet2(1,:)              = 0;
colormap(jet2)
```

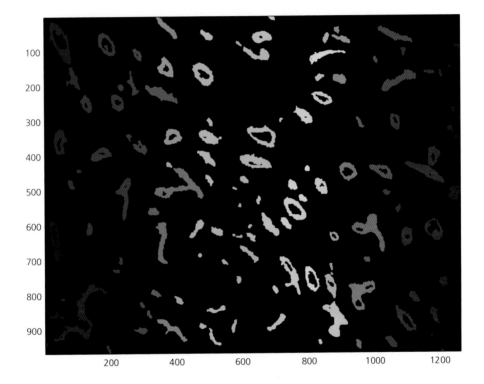

We can now change the order of the remaining rows to alternate the colours. The colormaps are normally matrices with 64 rows (which describe a range of shades) and 3 columns (for the individual red, green and blue components). Therefore, to change the order of the rows randomly we can create a matrix with random numbers between 0 and 1, sort that matrix and use the original order of the random elements to rearrange the colour map.

```
[index1,index2]        = sort(rand(64,1));
jet3                   = jet;
jet3                   = jet3(index2,:);
jet3(1,:)              = 0;
colormap(jet3)
```

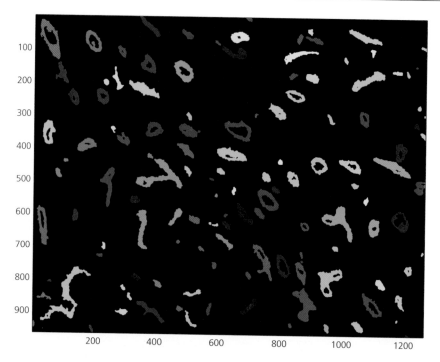

It is even possible to assign colours at random, for instance:

```
jet4                    = rand(64,3);
colormap(jet4)
```

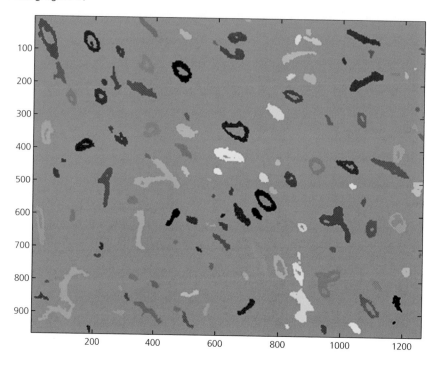

In some cases, random colours may not be visually attractive.

The easiest way to extract measurements from labelled elements is to use the command 'regionprops', which was introduced in Recipe 6. Let's explore more measurements that can be relevant to vessels:
- 'Area', number of pixels that form an object
- 'Perimeter', number of pixels in the boundary between the object and the background
- 'Eccentricity', a measurement of how circular (0) or linear (1) an object is
- 'EquivDiameter', the diameter of a circle with the same area as the object
- 'Euler number', number of holes inside an object
- 'Major / Minor axis length', length along a given axis
- 'Orientation', the angle at which the object is orientated
  The following images will illustrate some of these measurements:

```
dataIn_Properties  = regionprops(dataIn_Labelled,'Area',...
    'Perimeter','Centroid','Eccentricity','Euler',...
    'EquivDiameter','MajorAxisLength','MinorAxisLength',...
    'Orientation');


largeCells          = ismember(dataIn_Labelled,...
                      find([dataIn_Properties.Area]>800));
elongatedCells      = ismember(dataIn_Labelled,...
                      find([dataIn_Properties.Eccentricity]...
                      >0.9)  );
vesselsWithHoles    = ismember(dataIn_Labelled,...
                      find([dataIn_Properties.EulerNumber]<1));
vesselsOrientedLeft= ismember(dataIn_Labelled,...
                      find([dataIn_Properties.Orientation]...
                      <pi/2));

figure
subplot(221)
imagesc(data_channel2+largeCells)
title('Large and small vessels')

subplot(222)
imagesc(data_channel2+elongatedCells)
title('Round and elongated vessels')

subplot(223)
imagesc(data_channel2+vesselsWithHoles)
title('Cells with and without holes')

subplot(224)
imagesc(data_channel2+vesselsOrientedLeft )
title('Cells with different orientations')
colormap(gray)
```

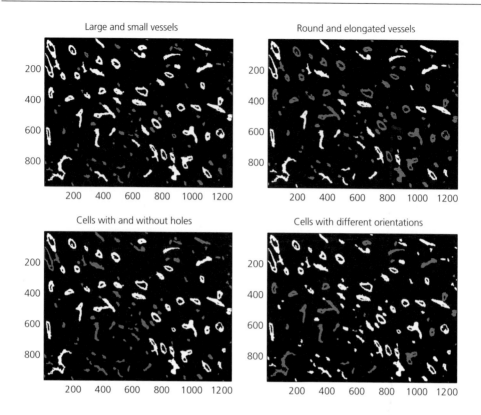

To extract other measurements we can use other operations and combine some of the measurements. For instance, in some cases it may be interesting to measure the size of the lumen of a microvessel and compare it to the size of the vessel itself. If we start with the vessels that have holes (i.e. a lumen) we can measure their areas, then fill the holes and measure the areas again:

```
vesselsWithHoles_filled        = imfill(vesselsWithHoles,'holes');

area_vesselsWithHoles          = regionprops...
                                 (vesselsWithHoles,'Area',...
                                 'Centroid');
area_vesselsWithHoles_filled = regionprops...
                                 (vesselsWithHoles_filled,...
                                 'Area');
```

The areas previously obtained are the absolute measurements in pixels. It may be more interesting to calculate the ratio of the areas, that is a ratio that tends to 1 corresponds to a vessel with a very small lumen, and a ratio that tends to 0 corresponds to a vessel with a very thin wall and a very large lumen.

```
ratioLumen_Vessel = [area_vesselsWithHoles.Area] ./...
                    [area_vesselsWithHoles_filled.Area];
```

```
figure
plot(ratioLumen_Vessel,'r-o')
grid on
axis tight
ylabel('Ratio of area of the vessel wall to the full vessel')
```

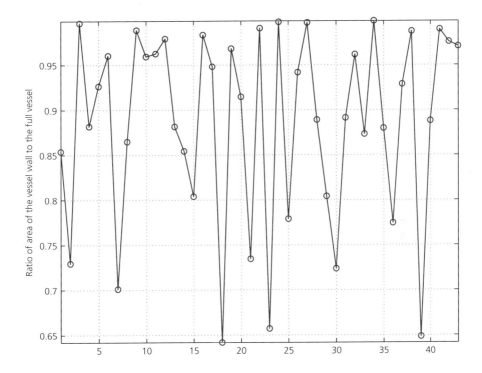

From the previous plot we can see that the majority of the vessels are solid with small lumen relative to the size of the vessel. It may be more useful to observe the vessels with their corresponding ratios. We can plot the vessels and then add a text next to them using the command 'text' and passing the values of the centroids and the ratio previously calculated. Notice that the ratio is a number, and then to be displayed as a text we need the command 'num2str' to convert to a string.

```
figure

imagesc(vesselsWithHoles)

for k=1:size(area_vesselsWithHoles,1)
    text(area_vesselsWithHoles(k).Centroid(1),...
        area_vesselsWithHoles(k).Centroid(2),...
        num2str(ratioLumen_Vessel(k),2),'color','r');
end
colormap(gray)
```

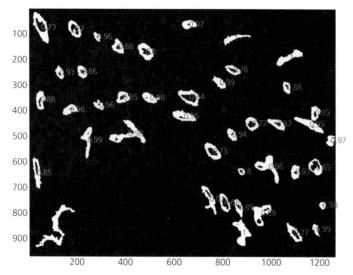

Another interesting measurement corresponds to the thickness of the walls of the vessels. An average measurement could be obtained from the area and the perimeter, but it may be the case that we are interested, not in the average but the maximum or minimum thickness. To do this we can use a few intermediate steps for which we need other useful functions.

The first function can easily calculate the distance of an image in several ways. The function 'bwdist' calculates different distance measurements between objects in an image. For example, to calculate the distance of every background pixel to its nearest vessel we use 'bwdist' in the following way:

```
distanceToVessels            = bwdist(vesselsWithHoles);

figure
imagesc(distanceToVessels)
colorbar
colormap(gray.^0.5)
```

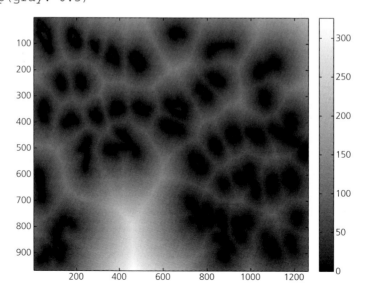

The image of the distance map displays the distance as a colour-coded map. Notice that we have manipulated the 'gray' colour map by raising it to the power of 0.5, which is the equivalent of a square root. With this manipulation we have given a higher emphasis to the lower intensities.

In this example, the point that is furthest away from a vessel is in the bottom of the region and corresponds to a distance of 325 pixels. In the same way, we can calculate the distances inside the vessels by inverting the image before calculating the distance:

```
distanceToBackground          = bwdist(1-vesselsWithHoles);
```

```
figure
imagesc(distanceToBackground)
colorbar
colormap(gray.^0.5)
```

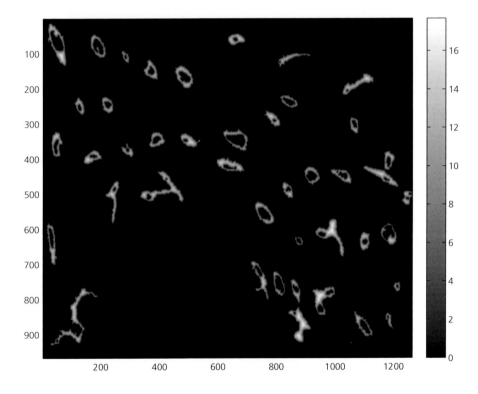

The image now shows that the distance from each point inside the vessels to the background, which includes the lumen, is between 1 and 18 pixels. To observe the detail we can zoom in to view some interesting vessels.

```
figure
imagesc(distanceToBackground(310:380,340:560))
colorbar
colormap(gray.^0.5)
```

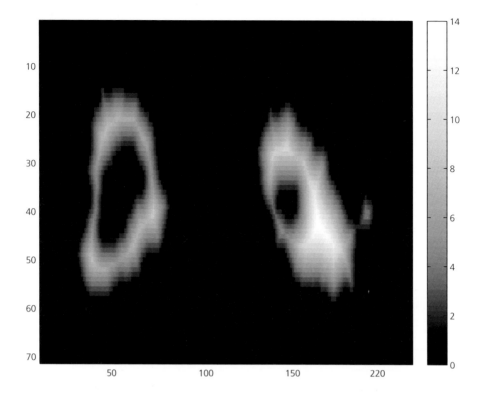

The vessel on the left is fairly uniform in thickness, but the vessel on the right has a very thin region and a thick section on the right. To investigate how the thickness is distributed along the vessel we need one extra step, that is, we need to draw a line along the vessel in the centre of the wall, or somewhere very close to the centre. To do this we can make use of one of many "morphological operators" that can modify the objects with some simple rules. Two of the most common morphological operators are used to dilate and erode objects:

```
vesselsEroded            = bwmorph(vesselsWithHoles,'erode');
vesselsDilated           = bwmorph(vesselsWithHoles,'dilate');
```

```
figure
subplot(121)
imagesc(vesselsEroded)
subplot(122)
imagesc(vesselsDilated)
colormap(gray.^0.5)
```

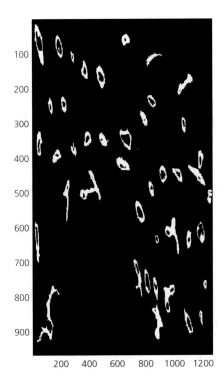 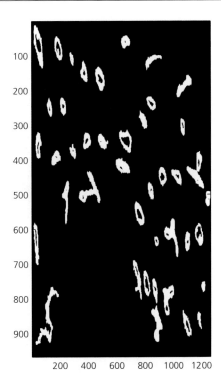

As the name implies, erosion will remove the outer pixels of an element. In the example an operator of size 3 × 3 is used, but many sizes and shapes can be used, and dilation will add pixels to the boundaries of every element. These operations can be used sequentially many times and with different conditions, and they can produce interesting results. One particularly interesting case is obtained when the pixels or the objects are removed sequentially, then there are several possibilities: keep removing until the object disappears, stop removing when the object becomes a single point, stop removing when the object is a line of 1 pixel width. The command 'bwmorph' provides several algorithms that apply these possibilities, namely skeletonisation, thinning and shrinking, which produce different results as we can see:

```
vessels_skeleton    = bwmorph(vesselsWithHoles,'skel','inf');
vessels_thin        = bwmorph(vesselsWithHoles,'thin','inf');
vessels_shrink      = bwmorph(vesselsWithHoles,'shrink','inf');
```

```
figure
subplot(131)
imagesc(vesselsWithHoles(320:370,360:420)+...
        vessels_skeleton(320:370,360:420))
title('Skeleton')
subplot(132)
```

```
imagesc(vesselsWithHoles(320:370,360:420)+...
        vessels_thin(320:370,360:420))
title('Thin')
subplot(133)
imagesc(vesselsWithHoles(320:370,360:420)+...
        vessels_shrink(320:370,360:420))
colormap(gray)
title('Shrink')
```

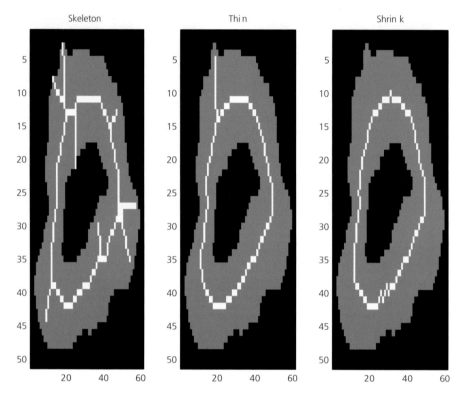

For our particular problem, we are interested in obtaining a ring or centre line around every vessel to investigate the thickness, so shrinking seems to be the most suitable algorithm. If we combine the distances previously obtained with the centre lines we can obtain the thickness around the vessel at every point. To visualise this, we can display the distance map for the previous example, first removing the centre line, and second displaying only those values of distance along the centre line.

```
vessels_centreline = bwmorph(vesselsWithHoles,'shrink','inf');

figure
subplot(121)

imagesc(distanceToBackground(320:370,360:420).*...
    (1-vessels_centreline(320:370,360:420)))
subplot(122)
```

```
imagesc(distanceToBackground(320:370,360:420).*...
    vessels_centreline(320:370,360:420))
colormap(gray)
```

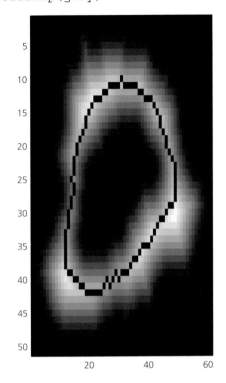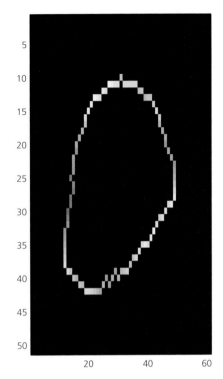

Finally to obtain the measurements for every vessel of the figure, we need to proceed iteratively over each of them. To do this, we need to individually label the centre lines first:

```
[vessels_centreline_labelled,numVessels] ...
    = bwlabel(vessels_centreline);
```

Now we iterate over every vessel; we select the pixels of the distance map by addressing that matrix with the labelled centre lines. Addressing a matrix is explained in Chapters 1–3.

```
for counterVessels = 1: numVessels
    distanceOverCentreLine        = ...
        distanceToBackground...
          (vessels_centreline_labelled==counterVessels);
    maxThickness(counterVessels)  = ...
        max(distanceOverCentreLine);
    minThickness(counterVessels)  = ...
        min(distanceOverCentreLine);
    meanThickness(counterVessels) = ...
        mean(distanceOverCentreLine);
end
```

```
figure

plot((1:numVessels),maxThickness,'b--o',...
    (1:numVessels),minThickness,'k--d',...
    (1:numVessels),meanThickness,'r-x')
grid on
xlabel('Vessels');
ylabel('Vessel Thickness [pixels]')

axis tight
```

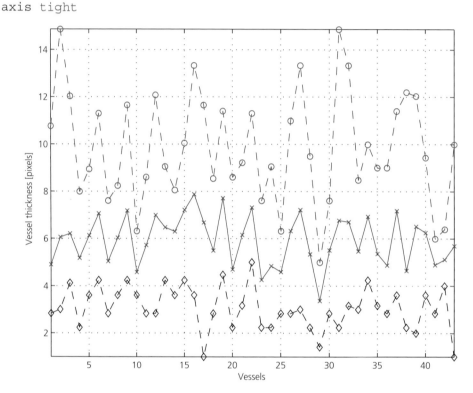

Once we have a series of instructions that provide us with a relevant result, such as the thickness of vessel walls, we can save it as an m-file. Then we can use it to compare different images. For example, we can have two populations, one of cells that have been treated with a certain compound and we want to compare it against a control. With this algorithm, the comparison can be performed with a single "loop" that reads every image, compiles the measurements and stores the results in a matrix to which we will test, for instance with 'ttest2'.

More complex morphological operations like linking or splitting objects are beyond the scope of this book, see reference Reyes-Aldasoro et al. 2011 for more details.

A final word of caution; all the measurements have been acquired and calculated from the pixels and intensity values of the images. The measurements that are reported in publications on the other hand are normally not in pixels, but in units of the International System of Units, like metres or seconds.

There are two common procedures to follow. The first is to divide a result by the same unit to obtain a relative measurement, such as the per cent of an area, which then is not dependent of the scale or the equipment used to obtain the data. The second is to normalise the measurements, e.g. to translate from pixels to a measure of distance. To do this, we need to know the equivalence between the dimensions of the image and the dimensions of the area it covers. A common procedure to know the equivalence is to use a "stage micrometre", which is a very fine and small ruler. When the samples are acquired with a microscope, an extra image should be taken with the stage micrometre with exactly the same conditions of the experiment. The image can then be used to calibrate the ratio of pixels to micrometres. Then, in Matlab, the known distance (in micrometres for instance) from the stage can be measured (in pixels) with 'imtool' and that ratio can be used to calibrate all the results into micrometres, or any other unit.

References

Otsu, Nobuyuki. 1979. "A Threshold Selection Method from Gray-Level Histograms." IEEE Transactions on Systems, Man and Cybernetics 9, No. 1 (January): 62–66. doi:10.1109/TSMC.1979.4310076.

Reyes-Aldasoro, Constantino Carlos, Williams, Leigh J., Akerman, Simon, Kanthou, Chryso, Tozer, Gillian M. 2011. "An automatic algorithm for the segmentation and morphological analysis of microvessels in immunostained histological tumour sections." Journal of Microscopy 242 (3), 262–278. doi: 10.1111/j.1365-2818.2010.03464.x.

CHAPTER 8

Creating Publication-Quality Figures from Matlab

One crucial part of any research is the generation of publication-quality figures that can be submitted to journals. In the previous sections of this book, we have developed skills to analyse and display a great variety of images. In this section, we will combine those skills and apply them to generate publication-quality figures.

There are many ways of generating and manipulating figures in Matlab. In Chapter 1 we covered the basic way to print an image to a file through the command 'print'. This command requires several arguments to print an image. Some of the most important arguments define the following: (a) the format of the output image; '-djpeg' for jpegs is common for low resolution images to be placed on web pages, '-dtiff' for tiffs is one of the most used by journals as it has better quality than jpegs, (b) the resolution, '-r100' corresponds to 100 dots per inch and is useful for low quality image to be used in web pages, '-r300' is common for high quality colour figures, '-r500' or higher can be required sometimes for half-tones and (c) the name of the file, like 'nameOfFile.jpg'.

It is possible to generate an image and then type print from the Command Window with the corresponding details. For quick printing, it is better to write a shortcut (described in Chapter 1), which can be used to save figures in files with one click. Another possibility is once a figure has been displayed, we can use the FILE > SAVE AS, or FILE > PRINT options from the menu of the figure and select the format required.

However, high quality figures such as those to be submitted to journals normally require the addition of labels, text, legends, arrows or other markers. Additionally, it is common to modify an image several times before it reaches the state at which it can be submitted. Therefore, the best way to handle a high quality figure is by writing a specific script file for it. This script file will then contain all the commands necessary to display the image plus any markers or text. If the image changes, or requires additions, it is easy then to modify the file and generate a new figure just by executing the script file as many times as necessary without having to repeat the process, step by step, by which the figure was created.

Before we generate high quality images we will explore the characteristics of figures and how to manipulate them. Then we will illustrate the process of high quality image generation with three examples; the files for these images will be Figure1_JournalA.m, Figure2_JournalB.m and Figure3_JournalC.m.

Biomedical Image Analysis Recipes in MATLAB®: For Life Scientists and Engineers, First Edition.
Constantino Carlos Reyes-Aldasoro.

8.1 Recipe 1: Modifying the Characteristics of the Figures

In some cases it is useful to change the size and position of a figure on the screen. This can be done manually by moving the figure with the mouse or another pointing device. However, it can also be done automatically, and it may be convenient in cases where many images are being manipulated, or when user interfaces are used. In those cases, it is better to use commands rather that resizing manually. To reposition and resize a figure, we first need to refer specifically to the figure we want. In Matlab each "element" such as a figure, an image or an axis is defined by its `'handle'`. The `'handle'` contains a large amount of information that defines the element or object itself. For the time being, suffice it to say that a figure has a position defined by the bottom-left coordinates of the window and the horizontal and vertical dimensions that it spans towards the right and up. To obtain the handle of the current figure (the last one to be used) we use the command `'gcf'`, which stands for "get current figure". Then, we can "get" or "set" its properties. For example, to get the current position of the window we use the command `'get'` with arguments `'gcf'` and `'Position'` like this:

```
load('immunohisto_CD31.mat')
imagesc(immunohisto_CD31);
get(gcf, 'Position')
```

```
        ans =

            560    528    560    420
```

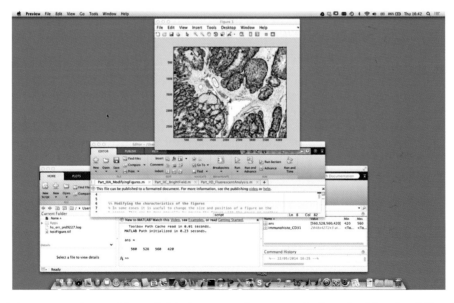

We can see the position of this image at the centre is described by the following values [440 378 560 420], which correspond to [bottom, left, width, height].

To modify the position, we need to change the values stored in the handle. To do this, we use the command 'set', again with 'gcf' and 'Position' and in addition a matrix with four values, for example:

```
set(gcf, 'Position', [10 40 800 310]);
```

The new coordinates have placed the figure in the bottom-left corner, i.e. 40 screen lines from the bottom and 10 screen lines from the left.

The position values are absolute values that depend on the size of the screen on each computer. To avoid problems of having figures that span beyond the edges of the screen, it is easy to capture the screen size and use values relative to it. First we save the size of the screen with the command 'get' in the variable 'currentScreenSize'.

```
currentScreenSize                         = get(0, 'ScreenSize')


    currentScreenSize =

            1            1          1680          1050
```

To create a figure that covers the whole screen we could resize like this:

```
set(gcf, 'Position', [1 1 currentScreenSize(3)...
    currentScreenSize(4)])
```

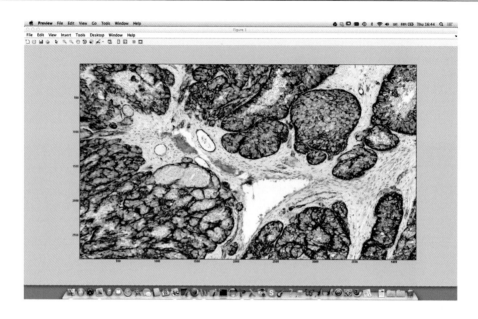

And a smaller version, at the centre could be obtained like this:

```
set(gcf, 'Position', [0.25*currentScreenSize(3) ...
    0.25*currentScreenSize(4) ...
    0.5*currentScreenSize(3) 0.5*currentScreenSize(4)])
```

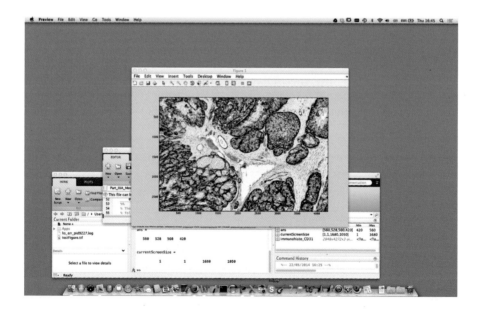

There are a few interesting modifications that can be performed by changing the following characteristics of a figure. First we can change the background colour from the default value that is a light grey ([0.8 0.8 0.8] in RGB), to a dark blue:

```
set(gcf, 'color', [0 0 0.5])
```

We can also remove the title of the figure (i.e. "Figure 1") and the menubar with the tabs ("File", "Edit", "View", etc.) as well as the icons for "New file", "Open file", "Save", "Print", etc.

```
set(gcf, 'numberTitle', 'off')
set(gcf, 'menuBar', 'none')
```

It is possible to modify several characteristics of a figure in one single instruction line by concatenating the characteristic and the new value. For example, we can return the image to its original conditions like this:

```
set(gcf, 'numberTitle', 'on', 'menuBar', 'figure', 'color', ...
    0.8*[1 1 1])
```

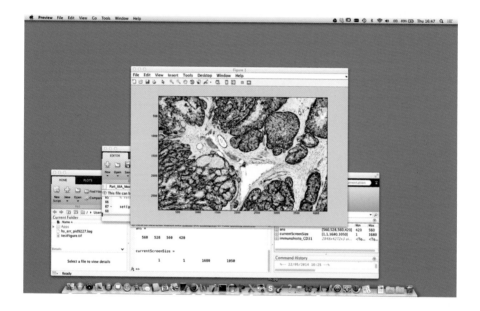

All the previous modifications were performed on the figure, which is the window that holds one or more "axes" where the graphs are placed. We can also modify the 'axes', that is, the image is placed in one axes inside the figure. Instead of using the command 'gcf' to address the figure, we can use the command 'gca' which stands for "get current axis". Alternatively, we can save the handles of an axes in a variable. In some previous recipes we had used multiple axes without saving their handles in variables. To save the handles, we simply write the name of the variable before the command 'subplot' like this:

```
figure
handlesAxis_2    = subplot(122);
imagesc(immunohisto_CD31(600:800, 2800:3100, :));


handlesAxis_1    = subplot(121);
imagesc(immunohisto_CD31);
```

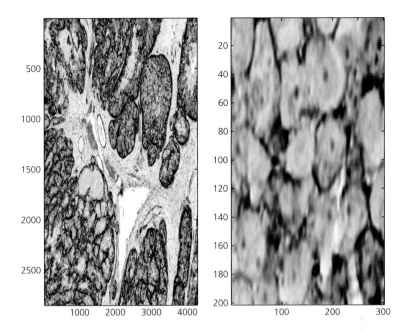

Notice that we can use the command 'subplot' in any order that we want.

We can now address each axis separately, by using the command 'set' with the corresponding handle. We have now full control of our figure. For example, we can resize to any arbitrary position and clear the scales from one of them with three commands:

```
set(handlesAxis_1, 'position', [0.07   0.15   0.58   0.80 ]);
set(handlesAxis_2, 'position', [0.70   0.40   0.27   0.50 ]);
set(handlesAxis_2, 'visible', 'off')
```

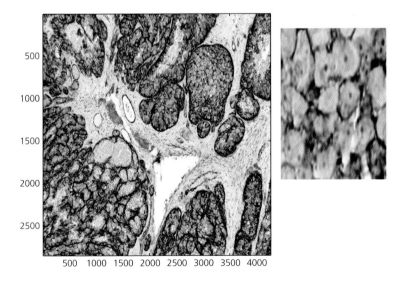

We have now two images of different size, one large and one small. In the small one we remove the "ticks", which we will explain below. Notice that the coordinates of the position are in relative terms to the size of the figure, and not in absolute terms as was the case of the figure.

One very important property of the axes is the values and marks that are placed along the vertical and horizontal axes. In the language of Matlab, the positions are called "ticks", and the characters for each tick are called "labels", thus we have 'xtick' and 'xticklabel', 'ytick' and 'yticklabel' and 'ztick' and 'zticklabel'. The ticks are spaced along both axes proportionally and the labels correspond to the number of pixels. For example, let's increase the number of ticks in the vertical dimension and reduce them in the horizontal dimension:

```
set(handlesAxis_1,'xtick',1000:1000:4000,'ytick', ...
    250:250:3000);
```

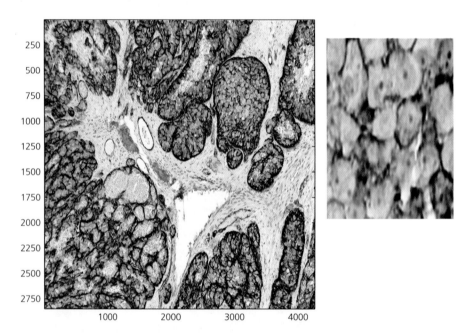

One instance in which we would be interested in modifying the labels of the ticks is when we know the size of the pixels and we want to display the data in absolute terms, such as micrometres for instance. Let's assume that every pixel measures 0.25 micrometres, thus the labels have to be scaled by this value. We can add a grid to help visualise dimensions:

```
set(handlesAxis_1, 'xticklabel', (1000:1000:4000)*0.25, ...
                   'yticklabel', (250:250:3000)*0.25);
handlesHorizontalLabel = xlabel...
   ('[dimensions in micrometers]');
grid on
```

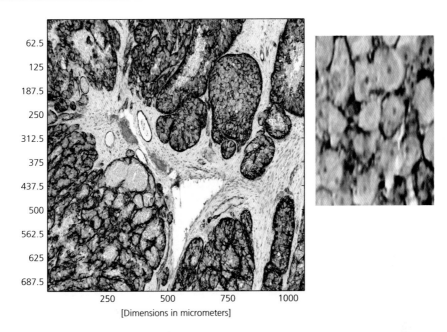

[Dimensions in micrometers]

Notice that we have also used a handle for the label of the horizontal axis. Actually, everything that appears in a figure has a corresponding handle that we can use to get or set parameters. Now that we have the handle of the label, we can use it to modify the text of the label by setting one of its parameters. To add special characters like the Greek letter μ, we use LaTeX-style equation formatting:

```
set(handlesHorizontalLabel, 'string', '[dimensions in \mum]',...
    'fontsize', 16)
```

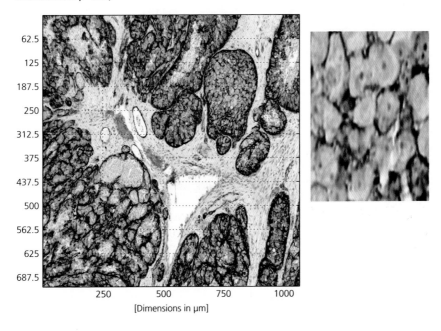

[Dimensions in μm]

LaTeX is a markup language for the formatting and editing of documents that is widely used in scientific fields. LaTeX uses a typesetting system called TeX that is used to format mathematical symbols, formulae and different typefaces. LaTeX originated as a tool for mathematicians, physicist and computer scientists, and it is still widely used.

We can also change the font of the numbers of the labels; the default font is 10-point "Helvetica". To change it to 12-point "Arial" and italics we use the command 'set' like this:

```
set(handlesAxis_1, 'fontname', 'arial', 'fontsize', 12, ...
    'fontangle', 'italic')
```

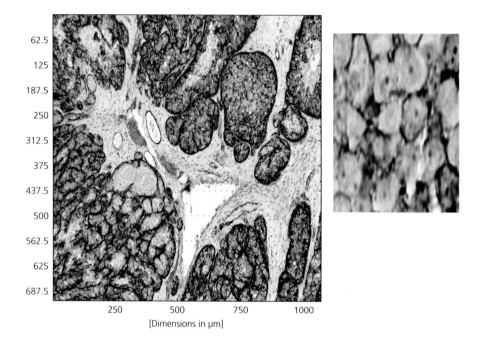

[Dimensions in μm]

This section has demonstrated how easy it is to modify figures, axis and labels. These changes will be explored in more detail in the next recipes.

8.2   Recipe 2: Numerous Plots in One Figure

In some journals such as "Nature" and its affiliates, the figures are normally subdivided into several panels, each of which contains one or more graphs or plots. One way to generate these figures is to create each section separately, save as different files and then merge them within a graphics-oriented software. However, with Matlab it is easy to generate a figure with several axes and display a variety of graphs in each of them. For this example, we will simulate a longitudinal study of three separate experiments; for each experiment, two different measurements were acquired. We wish to display the results of the experiments in several different graphs, each of which reveals something interesting of the data.

The longitudinal experiments will be simulated by a series of random values with different means and variances: one series will increase values over time, another series will decrease values over time and the third one will remain constant. The following lines will create the data sets:

```
experiment_1(:,1)     = 1*randn(1000,1);
experiment_1(:,2)     = 4*randn(1000,1);
experiment_2(:,1)     = -10+1.8*randn(1000,1) ...
                        +linspace(-3,3,1000)';
experiment_2(:,2)     = -10+3*randn(1000,1);
experiment_3(:,1)     = 6+3.7*randn(1000,1) ...
                        +linspace(3,-3,1000)';
experiment_3(:,2)     = 11+1.3*randn(1000,1);
```

The first graph will be a plot of the individual values against time for one of the measurements. This type of graph can be obtained using the command 'plot'. Before that, we will create a figure for all the graphs and will extend the figure so that it is in landscape format.

```
figure
set(gcf,'position',[ 100   400   1000 500])
```

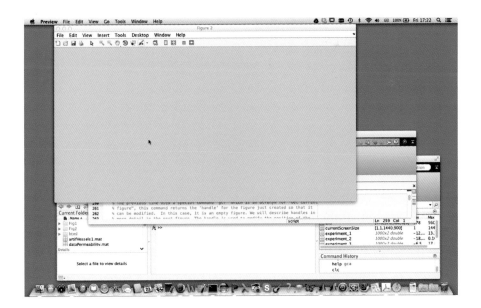

The previous line used the command 'gcf', which is an acronym of "Get current figure". This command returns the 'handle' for the figure just created so that it can be modified. In this case, it is an empty figure. The handle is used to modify the position of the figure so that it is longer (1000 pixels) than taller (500 pixels).

Next, we will create the first axis of the figure in which we will display the values against time for the first measurement of the three experiments. For this experiment, we will use the command 'subplot' to arrange the different graphs on the figure. The command 'subplot' takes three parameters as input, sometimes as three consecutive numbers 'subplot(236)' or with the numbers separated by commas like 'subplot(2,3,6)'. This last option is necessary when numbers are greater than 9. The first number corresponds to the rows into which the figure will be subdivided, the second corresponds to the number of columns and the third one indicates the current subplot, i.e. 6 corresponds to the bottom right. In the next example we will manipulate subplots and their positions.

For this example, we will first subdivide the figure in two rows and two columns and display the first plot on the top left.

Since we want to display several things on the axis, we start indicating the figure to hold everything after the first plot, as the default is to refresh the figure for each plot.

```
handle1 = subplot(221);
hold on
plot((1:1000),experiment_1(:,1),'b',...
     (1:1000),experiment_2(:,1),'r',...
     (1:1000),experiment_3(:,1),'k')
```

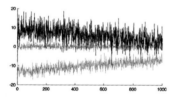

We now assume that there is a region of interest over a certain moment in time. We can overlay a box over the region of interest to highlight that region, and we will zoom into that region by displaying those values in another axis. The region of interest will span the time points (800,900) and the range of values (−15,5). To draw a square we can simply use the command 'plot' again with 5 points in the horizontal and vertical dimensions and a thick black line:

```
plot([800 900 900 800 800],[5 5 −15 −15 5],'k','linewidth',4)
```

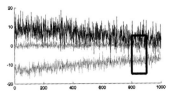

We need five points to draw the four segments sequentially. Finally, we can add a title, labels and a legend to describe the graph as well as setting the limits for the axis and a grid:

```
title('(a)','fontsize',20)
xlabel('Time [sec]','fontsize',14)
ylabel('Measurement 1 [a.u.]','fontsize',14)
legend('Experiment 1','Experiment 2','Experiment 3', ...
       'Location','Northwest')
axis([1 1000 -20 20])
grid on
```

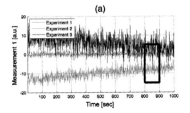

The region of interest will be displayed as a small blow up to the right of the previous graph. For this purpose we will subdivide the figure into 3 rows and 6 columns and use only one of those options. We will display the values with the command 'plot', exactly the same as before, then we can use the command 'axis' to blow up the region between (800,900) in the time and (−15,5) in the measurement.

```
handle2 = subplot(3,6,4);
plot((1:1000),experiment_1(:,1),'b',...
     (1:1000),experiment_2(:,1),'r',...
     (1:1000),experiment_3(:,1),'k')
axis([800 900 -15 5])
```

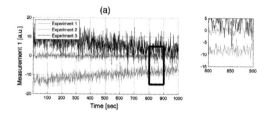

To give a clear impression that the new plot corresponds to a zoom of the region of interest, we will add two dashed lines to indicate from where the zoom region has been obtained. The lines are added with the command 'annotation' with which we can add lines, arrows, text, boxes or ellipses to our figures.

```
line1 = annotation(gcf,'line',[0.424     0.535], ...
        [0.79     0.9237],...
    'linewidth',4,'linestyle','--');
line2 = annotation(gcf,'line',[0.424     0.535], ...
        [0.6250     0.7100],...
    'linewidth',4,'linestyle','--');
```

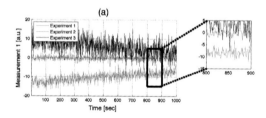

In the next three axes we will explore the distribution of the values of the experiments with useful graphics common in descriptive statistics.

First we will add a new axis to display a boxplots for the second measurement. Boxplots are a very useful statistical tool to display a distribution of values; they are specially useful when we want to compare the statistical difference between groups. In this case, we will use the command 'boxplot' with two arguments: the first argument is a matrix with 3 columns, each column corresponds to the values of one experiment and the second argument is a 'cell' with the three names that correspond to the experiments.

In Matlab, round brackets or parenthesis "( )" are used to indicate an order of execution of commands, e.g. '(a+b)*(d-(e*f))' or to indicate that we are passing parameters to a certain function, e.g. 'sqrt(4)' or to address a matrix as has been previously explained. Square brackets "[ ]" on the other hand indicate that we are describing a matrix, e.g. '[1 2 3; 4 5 6]'. These matrices are numerical and follow all the rules of linear algebra for addition, multiplication, etc. Finally, curly brackets "{ }" are used to create 'cells'. Cells are different from matrices in that cells can have many different data types in each cell. Indeed one element of a cell can contain a matrix, another a string, another one number. Look at this example:

```
a{1} = 'This is a string';
a{2} = [1 2 3; 4 5 6];
a{3} = randn(256)>0.5;
a{4} = [];
```

disp (a)

'This is a string'   [2×3 double]   [256×256 logical]   []

The position of this plot in the figure will correspond to a subplot of 2 rows and 3 columns to be on the right of the previous graphs.

```
handle3 = subplot(2,3,3);
boxplot([experiment_1(:,2),experiment_2(:,2), ...
   experiment_3(:,2)],{'Exp 1','Exp 2','Exp 3'})
axis([0.5 3.5 -20 20])
grid on
ylabel('Measurement 2 [a.u.]','fontsize',14)
title('(b)','fontsize',20)
```

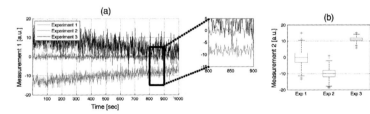

The boxplots indicate the distribution of values of the experiments; whilst the first experiment is centred at 0, the second is formed by negative values and the third one by positive values. The red crosses correspond to values that are considered as outliers of the distribution. From the boxplots it is very easy to see if the values of measurements of the two experiments overlap or not.

A useful way to display more than one measurement in a single graph is to use "scatter plots" with one measurement displayed on the vertical axis and the other measurement displayed over the horizontal axis. We will now create another subplot to display the scatter plot underneath the first graph. We will again use 'plot', but instead of plotting the measurements against the time (or the order in which the appear in the matrix) we will plot one measurement against the other one:

```
handle4 = subplot(2,2,3);
handleScatter = plot(experiment_1(:,1), experiment_1(:,2), ...
                     'b.',...
                     experiment_2(:,1), experiment_2(:,2), ...
                     'r+',...
                     experiment_3(:,1), experiment_3(:,2), ...
                     'kx')
axis([-20 20 -20 20])
grid on
xlabel('Measurement 1 [a.u.]','fontsize',14)
ylabel('Measurement 2 [a.u.]','fontsize',14)
legend('Experiment 1','Experiment 2','Experiment 3', ...
       'Location','Northwest')
title('(c)','fontsize',20)


    handleScatter =

        678.0094
        679.0089
        680.0089
```

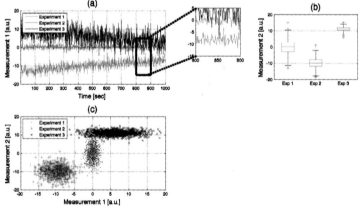

Notice that we are only using a marker (".", "+", "x") and no lines ("-", "--", "-.", ":"). When we display scatter plot it is better not to use a line, as the line connects between the points in the order that they are in the matrices, and this is not relevant for a scatter plot. Look at the scatter plot with the lines between points:

```
set(handleScatter,'linestyle','-')
```

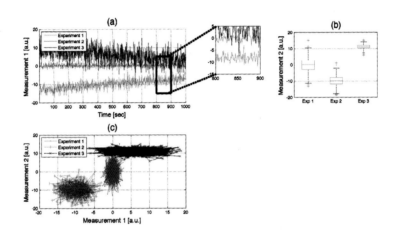

Finally we will display a 2D histogram with meshes for each experiment, again for one measurement against the other one. To obtain the 2D histograms, first we will calculate the 1D histograms like this:

```
experiment_1_x    = hist(experiment_1(:,1),(-20:20));
experiment_1_y    = hist(experiment_1(:,2),(-20:20));
experiment_2_x    = hist(experiment_2(:,1),(-20:20));
experiment_2_y    = hist(experiment_2(:,2),(-20:20));
experiment_3_x    = hist(experiment_3(:,1),(-20:20));
experiment_3_y    = hist(experiment_3(:,2),(-20:20));
```

Now we can multiply the 1D histograms to approximate the 2D histograms; we can do this multiplication at the same time that we pass them as an input argument to the command 'mesh'. We will use different colours to identify each distribution.

```
handle5 = subplot(2,2,4);
hold on
mesh((-20:20),(-20:20),experiment_1_x'*experiment_1_y, ...
     'edgecolor','b');
mesh((-20:20),(-20:20),experiment_2_x'*experiment_2_y, ...
     'edgecolor','r');
mesh((-20:20),(-20:20),experiment_3_x'*experiment_3_y, ...
     'edgecolor','k');
xlabel('Measurement 1 [a.u.]','fontsize',14)
ylabel('Measurement 2 [a.u.]','fontsize',14)
zlabel('Number of cases','fontsize',14)
```

```
title('(d)','fontsize',20)
view(40,60)
axis tight
grid on
```

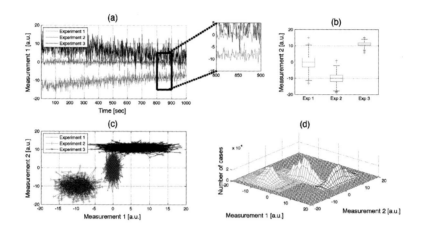

To clarify which mesh corresponds to each experiment, we will add annotations for the experiments with the corresponding colours:

```
annotation(gcf,'textbox',[0.71 0.34 0.0865 0.048],'String',...
          {'Experiment 1'},...
   'color','b','linestyle','none','fontsize',12);
annotation(gcf,'textbox',[0.62 0.31 0.0865 0.048],'String',...
          {'Experiment 2'},...
   'color','r','linestyle','none','fontsize',12);
annotation(gcf,'textbox',[0.78 0.31 0.0865 0.048],'String',...
          {'Experiment 3'},...
   'color','k','linestyle','none','fontsize',12);
```

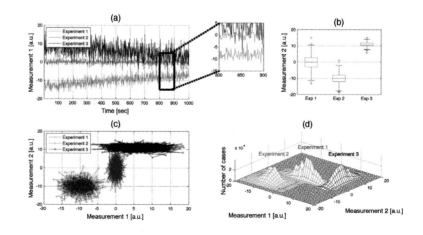

In some cases, you may prefer a different style of titles from the ones that we have displayed in this figure. You may increase the size or change the font, but you could easily remove those titles by setting the string of the title to an empty string like this:

```
subplot(handle1); title('')
subplot(handle2); title('')
subplot(handle3); title('')
subplot(handle4); title('')
subplot(handle5); title('')
```

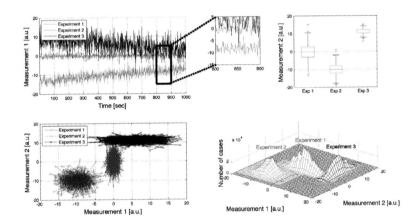

Now we can add text boxes that will identify the plots with the letters a, b, c, d. The letters will be in white and the box will have a black background. Notice that we can write the name of the colour (e.g. 'black', 'cyan') or we can use a shorthand notation (e.g. 'k', 'c').

```
annotation(gcf,'textbox',[0.02    0.74    0.06    0.12],...
    'String','(a)','fontsize',40,'color','black',
    'backgroundcolor','cyan')
annotation(gcf,'textbox',[0.92    0.74    0.06    0.12],...
    'String','(b)','fontsize',40,'color','k',
    'backgroundcolor','c')
annotation(gcf,'textbox',[0.02    0.32    0.06    0.12],...
    'String','(c)','fontsize',40,'color','k',
    'backgroundcolor','c')
annotation(gcf,'textbox',[0.92    0.32    0.06    0.12],...
    'String','(d)','fontsize',40,'color','k',
    'backgroundcolor','c')
```

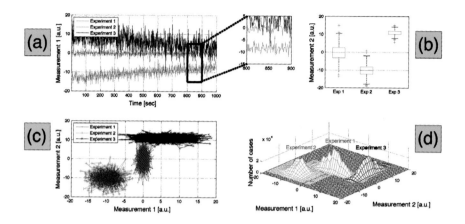

This example has shown how to generate an interesting figure with several graphs in a single figure. We could now copy all the commands and save them inside a single Matlab file. For instance, assuming that this will be the first figure of manuscript that will be submitted to a certain journal "A", the name of the file could be 'Figure1_JournalA.m'. We can save this file in the current working directory, and therefore if we want to generate the whole figure in a single command, we would type 'Figure1_JournalA' on the Command Window, and the figure would be generated following all the commands of the file. We could also add a print command (print('-dtiff','-r400','Figure1_JournalA.tif')) at the end of the file, and we could generate a 'tiff' file as well. The advantage of creating files to generate the images in this way is the fact that, if for some reason we need to modify something, say we discover that the outlier was an error at the moment of recording the values, we only need to change that value on the file, and we can rerun the file to generate the new image.

8.3 Recipe 3: Three-Dimensional Ribbons with Different Annotations

We will now illustrate different display techniques and annotations. Let's assume that we want to display the results from a proliferation experiment in which five different cell lines were observed under different conditions of temperature (t1, t2, t3, t4, t5) and a drug concentration (c1, c2, c3). The measurements obtained from the cells are (1) the number of cells observed in a certain region of interest and (2) the smallest distance between the cells of that region of interest.

The results were collected in two tables, one for each measurement, and each column corresponds to one cell line and the rows correspond to a combination of temperature and concentration in the following order: (t1,c1), (t1,c2), (t1,c3), (t2,c1), (t2,c2), (t2,c3), (t3,c1), etc. In some cases there were no cells detected in the region of interest and the distance was noted as 0. The results for the number of cells follow:

```
numberOfCells = [
    12      33      32       0       0      % t1, c1
    42      22      29      30       0      % t1, c2
    35      13      25      27       0      % t1, c3
    15       7      19      22      44      % t2, c1
     9       6      13      19      33      % t2, c2
     7       8       9      18      29      % t2, c3
     6       7       8      21      31      % t3, c1
     5       6       6      13      15      % t3, c2
     6       6       6       9      14      % t3, c3
     9       6       6       7      14      % t4, c1
     6       2       6       7      12      % t4, c2
     0       3       6       8      23      % t4, c3
     0       0       3       6       6      % t5, c1
     0       0       0       2       8      % t5, c2
     0       0       0       2      7];     % t5, c3
```

The corresponding results for the distances follow:

```
distanceBetweenCells=[
    2.1415    1.2053    0.5475         0         0
    1.1657    2.1710    2.2915    2.4322         0
    4.7249    3.9803    1.7069    1.0780         0
    5.4398    5.9109    3.7096    3.0363    0.3879
    6.3974    4.8951    5.8117    4.9891    1.6166
    6.7432    7.3670    4.9968    7.0731    3.6441
    7.4119    7.5507    6.0001    7.2854    2.9624
    6.9978    7.9976    6.9987    7.1830    2.9536
    7.3952    7.9980    7.2961    6.9799    4.7660
    7.6989    7.9990    7.3919    7.5719    5.6979
    8.0021    8.0013    7.5771    7.6667    5.9373
         0    7.8859    7.8469    7.8591    1.6473
         0         0    7.6813    7.8660    7.4237
         0         0         0    7.7183    7.8536
         0         0         0    8.0583    7.5536];
```

One simple way to display the results would be to plot them in separate figures like this:

```
figure
plot(numberOfCells)
figure
plot(distanceBetweenCells)
```

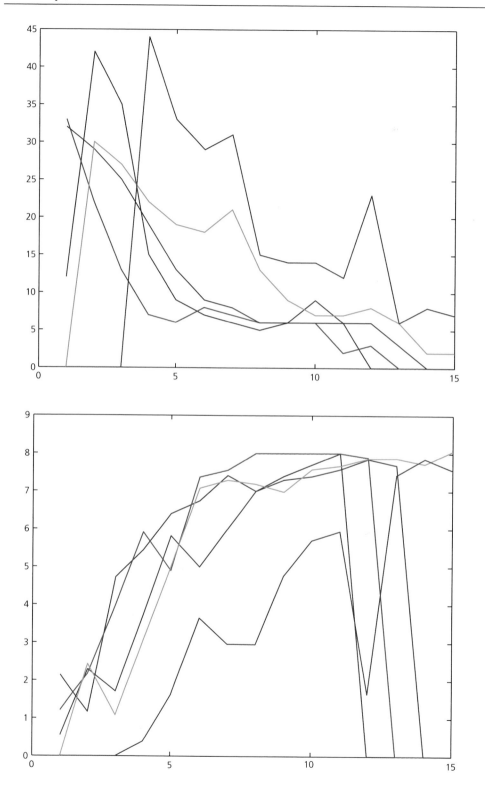

The general trends can be easily visualised from the plots. For one of the measurements, the number decreases with the increase of the axis, whilst the other, the distance increases with the axis. However, if the number of cell lines increases, this graph easily can become confusing, especially when one of the lines does not follow the trend in the same way as the others.

A better way to display these results would be to use a 3D graph that shows the variations of each cell line clearly, which we can achieve with the command 'ribbon'. For the modifications of this figure we will use extensively the 'handles' of all the elements in the figure.

```
handleFigure                        = figure;
handleAxes                          = axes;
handleRibbon                        = ribbon(numberOfCells);
```

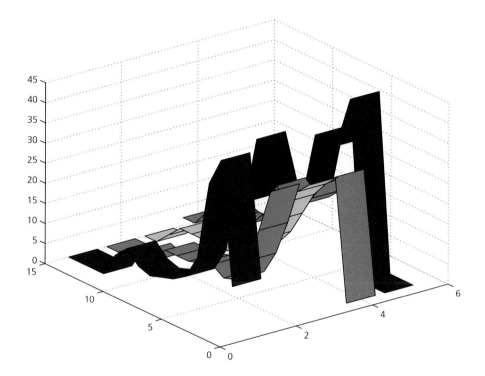

With the ribbons we can observe the variations of each cell line. Even if the number of lines grow, these can be better appreciated than with a simple 'plot'. The next step to improve the figure is to modify the labels to have meaningful text, rather than just numbers; here is where the handles are useful. First, we will modify the 'x' axis, which corresponds to the columns. The previous lines have assigned the handles to the variables. We could explore the properties of each by typing the following:

```
handleProperties_Figure             = get(handleFigure)
handleProperties_Axes               = get(handleAxes)
```

```
handleProperties_Figure =

              Alphamap: [1x64 double]
          BeingDeleted: 'off'
            BusyAction: 'queue'
         ButtonDownFcn: ''
              Children: 1.4040e+03
              Clipping: 'on'
       CloseRequestFcn: 'closereq'
                 Color: [0.8000 0.8000 0.8000]
              Colormap: [64x3 double]
             CreateFcn: ''
           CurrentAxes: 1.4040e+03
      CurrentCharacter: ''
         CurrentObject: []
          CurrentPoint: [0 0]
             DeleteFcn: ''
          DockControls: 'on'
              FileName: ''
      HandleVisibility: 'on'
               HitTest: 'on'
         IntegerHandle: 'on'
         Interruptible: 'on'
        InvertHardcopy: 'on'
           KeyPressFcn: ''
         KeyReleaseFcn: ''
               MenuBar: 'figure'
                  Name: ''
              NextPlot: 'add'
           NumberTitle: 'on'
        PaperOrientation: 'portrait'
          PaperPosition: [0.3607 2.5804 7.7690 5.8268]
       PaperPositionMode: 'manual'
             PaperSize: [8.5000 11]
             PaperType: 'usletter'
            PaperUnits: 'inches'
                Parent: 0
               Pointer: 'arrow'
       PointerShapeCData: [16x16 double]
     PointerShapeHotSpot: [1 1]
              Position: [560 528 560 420]
              Renderer: 'painters'
          RendererMode: 'auto'
                Resize: 'on'
             ResizeFcn: ''
              Selected: 'off'
    SelectionHighlight: 'on'
         SelectionType: 'normal'
                   Tag: ''
               ToolBar: 'auto'
                  Type: 'figure'
```

```
            UIContextMenu: []
                    Units: 'pixels'
                 UserData: []
                  Visible: 'on'
      WindowButtonDownFcn: ''
    WindowButtonMotionFcn: ''
        WindowButtonUpFcn: ''
        WindowKeyPressFcn: ''
      WindowKeyReleaseFcn: ''
      WindowScrollWheelFcn: ''
              WindowStyle: 'normal'
                 XDisplay: ''

handleProperties_Axes =

    ActivePositionProperty: 'outerposition'
                      ALim: [0.1000 10]
                  ALimMode: 'auto'
        AmbientLightColor: [1 1 1]
             BeingDeleted: 'off'
                      Box: 'off'
                BusyAction: 'queue'
            ButtonDownFcn: ''
           CameraPosition: [-24.3943 -81.7523 217.3557]
       CameraPositionMode: 'auto'
             CameraTarget: [3 7.5000 22.5000]
         CameraTargetMode: 'auto'
           CameraUpVector: [0 0 1]
       CameraUpVectorMode: 'auto'
          CameraViewAngle: 10.3396
      CameraViewAngleMode: 'auto'
                 Children: [5x1 double]
                     CLim: [1 5]
                 CLimMode: 'auto'
                 Clipping: 'on'
                    Color: [1 1 1]
               ColorOrder: [7x3 double]
                CreateFcn: ''
             CurrentPoint: [2x3 double]
          DataAspectRatio: [1 2.5000 7.5000]
      DataAspectRatioMode: 'auto'
                DeleteFcn: ''
                 DrawMode: 'normal'
                FontAngle: 'normal'
                 FontName: 'Helvetica'
                 FontSize: 10
                FontUnits: 'points'
               FontWeight: 'normal'
            GridLineStyle: ':'
         HandleVisibility: 'on'
                  HitTest: 'on'
```

```
        Interruptible: 'on'
               Layer: 'bottom'
       LineStyleOrder: '-'
           LineWidth: 0.5000
   MinorGridLineStyle: ':'
            NextPlot: 'replace'
       OuterPosition: [0 0 1 1]
              Parent: 5
    PlotBoxAspectRatio: [1 1 1]
PlotBoxAspectRatioMode: 'auto'
            Position: [0.1300 0.1100 0.7750 0.8150]
          Projection: 'orthographic'
            Selected: 'off'
   SelectionHighlight: 'on'
                 Tag: ''
             TickDir: 'out'
         TickDirMode: 'auto'
          TickLength: [0.0100 0.0250]
          TightInset: [0.0393 0.0381 0.0286 0.0024]
               Title: 1.4130e+03
                Type: 'axes'
       UIContextMenu: []
               Units: 'normalized'
            UserData: []
                View: [-37.5000 30]
             Visible: 'on'
       XAxisLocation: 'bottom'
              XColor: [0 0 0]
                XDir: 'normal'
               XGrid: 'on'
              XLabel: 1.4100e+03
                XLim: [0 6]
            XLimMode: 'auto'
          XMinorGrid: 'off'
          XMinorTick: 'off'
              XScale: 'linear'
               XTick: [0 2 4 6]
          XTickLabel: [4x1 char]
      XTickLabelMode: 'auto'
           XTickMode: 'auto'
       YAxisLocation: 'left'
              YColor: [0 0 0]
                YDir: 'normal'
               YGrid: 'on'
              YLabel: 1.4110e+03
                YLim: [0 15]
            YLimMode: 'auto'
          YMinorGrid: 'off'
          YMinorTick: 'off'
              YScale: 'linear'
               YTick: [0 5 10 15]
```

```
        YTickLabel:  [4x2 char]
    YTickLabelMode:  'auto'
         YTickMode:  'auto'
            ZColor:  [0 0 0]
              ZDir:  'normal'
             ZGrid:  'on'
            ZLabel:  1.4120e+03
              ZLim:  [0 45]
          ZLimMode:  'auto'
        ZMinorGrid:  'off'
        ZMinorTick:  'off'
            ZScale:  'linear'
             ZTick:  [0 5 10 15 20 25 30 35 40 45]
        ZTickLabel:  [10x2 char]
    ZTickLabelMode:  'auto'
         ZTickMode:  'auto'
```

Each of the properties of the handles has a specific use, however, not all of them are relevant at this moment, nor do they need to be modified. It is not necessary to save the handles into a variable, although it is recommended for complex images that will be further modified. It is always possible to refer to the "current" figure, or the "current" axis, which will be the last one to be displayed or the last one in which the user addressed with a mouse. In those cases we can use 'gcf', which is an acronym for "get current figure" or 'gca' for "get current axis". Therefore, the following two commands are equivalent:

```
set(gca,'xtick',(1:5))
set(handleAxes,'xtick',(1:5))
```

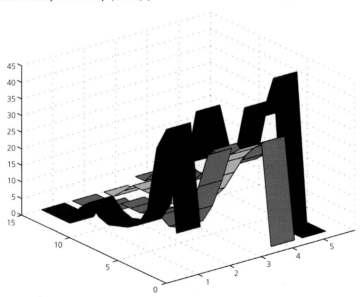

Matlab normally assigns the number of ticks along each axis automatically according to the dimension of the data and the size of the figure. The previous command(s) determined that the figure should have a 'tick' at every point of the 'x' axis, e.g. there will be five lines, one for every cell line along this dimension.

We will continue with the arrangement of the figure by removing extra space from the edges of the figure with the following command:

```
axis tight
```

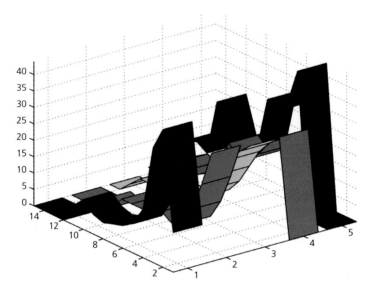

Next, we will change the labels of each axis. Instead of the numbers that correspond to rows and columns we will write a text relevant to the experiment. Since we determined that there would be 5 'ticks', we can now assign one label to each of them by setting the 'xticklabel' of 'handleAxes'. In the same way as before, we will enclose each label between single quotes, separated by commas and place all inside the curly braces of a 'cell':

```
set(handleAxes,'xticklabel',{'Line 1','Line 2','Line 3', ...
    'Line 4','Line 5'})
```

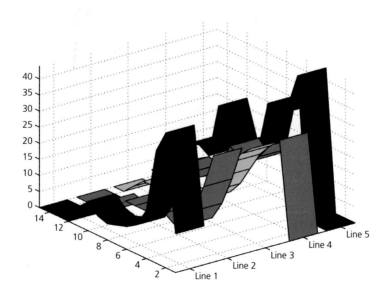

In the same way, we can now arrange the 'y' axis of the figure. In this case, we would like to display all the ticks along the axis, as each will have a special label. This is not necessary for other cases. For the example, ticks every 5 or 10 are sufficiently clear for the 'z' axis.

We thus set the 'y' axis to have 15 ticks, and for each tick, the legend will correspond to a temperature-concentration pair:

```
set(handleAxes,'ytick',(1:15))
set(handleAxes,'yticklabel',{'1,1','1,2','1,3',...
    '2,1','2,2','2,3',...
    '3,1','3,2','3,3',...
    '4,1','4,2','4,3',...
    '5,1','5,2','5,3'});
```

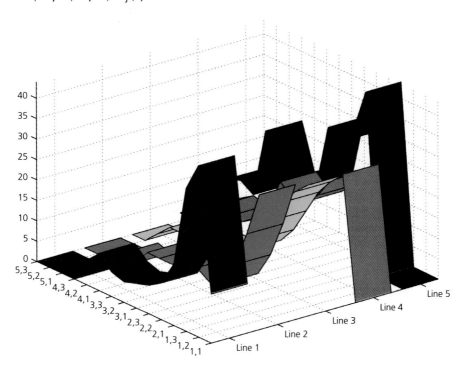

To complete the current figure, it is necessary to include labels along the three axes to specify what is expressed in each of them. Notice that we can control the properties, size, or kind of font, of each label separately. The default font is 10-point "Helvetica". We can exploit the fact that Matlab also interprets the formatting commands from 'Tex' and 'LaTeX', so we can include this kind of command to format the text. For instance we can change part of the string to italics with the command \it enclosed between curly braces {\it } or the command bold type like this {\bf}.

```
xlabel('Cell Lines','fontsize',15,'fontname','herculanum')
ylabel('Conditions ({\it Temp., Conc.})','fontsize',12,...
        'fontname','courier')
zlabel('Number of Cells ','fontsize',16,'fontname','times')
view(-100,30)
set(handleAxes,'fontsize',11)
```

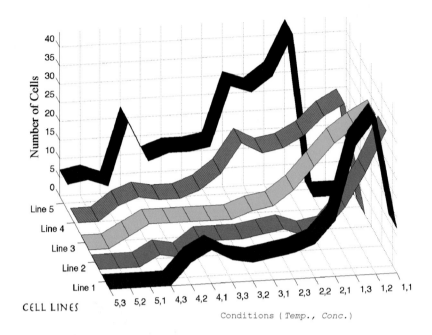

It is not recommended to mix the fonts of a figure; here it is used just to demonstrate that it is possible.

Now that we have displayed the number of cells with a ribbon, we realise that we would prefer to plot these values next to the corresponding distances in a single figure. One way to do that would be to start again, using `subplot (121)`, then type all the previous commands, then type `subplot (122)` for the second plot and type the commands again. With the use of handles, this is not necessary. We can display a second plot (technically, a second axes in the figure) and arrange the position of each axes next to each other. We first change the size of the whole figure so that we can have the two axes together in a horizontal arrangement.

```
set(handleFigure,'position',[543    600    904    356])
```

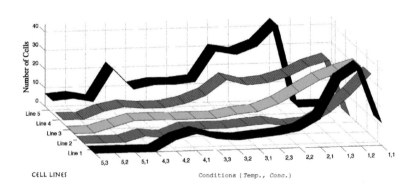

Notice that the position of the figure is specified in pixels; therefore, the previous values are suitable for a screen of (1440×990) pixels. Screens with other resolutions may need different values.

Next, we add the new axes to the figure. As before, we need to save the handles of the axes for further manipulation; we do this by typing:

```
handleAxes2                          = axes;
```

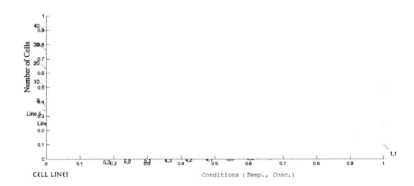

We can notice that the axes are overlapping and the previous plot is behind the new axes. We will reduce the sizes of both and send one to the left and the other to the right of the figure by changing their positions. Unlike figures, the positions of the axes are modified in relative units to the size of the figure, so we need to give values between 0 and 1. The position is defined from the bottom-left corner towards the upper right side as follows: [initial-column initial-row size-columns size-rows]. A suitable pair of positions for this example follows:

```
set(handleAxes ,'position',[0.0800   0.1500   0.3600   0.7000])
set(handleAxes2,'position',[0.5600   0.1500   0.3600   0.7000])
```

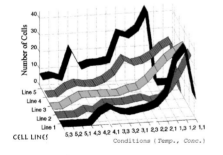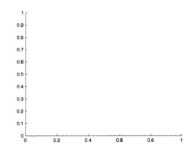

Now that the axes are in position we can display the distances between cells in the second axes with a ribbon as with the previous plot:

```
handleRibbon2                        = ribbon(distanceBetweenCells);
```

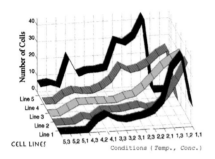 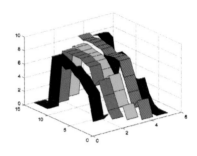

At this stage we notice something important. When recording the values of the experiment, we assigned a 0 for those cases where no cells were detected in the region of interest; however, this is not correct as a distance of 0 (or close to 0) in reality should represent a case where cells are very close to each other. For the cases with no cells the distance does not have a valid meaning and it would be better not to represent them in the graph. We can therefore remove these values from the ribbon by setting the 0 values to a special value called "not-a-number" or 'nan'. Another option could be to set the values to an upper limit value, which could be the size of the region of interest or an infinite value 'inf'. For this example we will set them to 'nan' and redraw in the following way:

```
distanceBetweenCells(distanceBetweenCells==0)=nan;
handleRibbon2                    = ribbon(distanceBetweenCells);
```

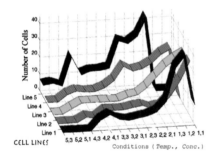 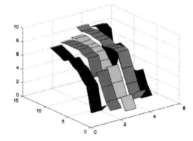

The ribbons now display only the locations with valid distances. Now with the ribbons in place, we can simply repeat all the commands that we used in the first plot:

```
set(handleAxes2,'xtick',(1:5))
axis tight

set(handleAxes2,'xticklabel',{'Line 1','Line 2','Line 3',...
    'Line 4','Line 5'})
set(handleAxes2,'ytick',(1:15))
set(handleAxes2,'yticklabel',{'1,1','1,2','1,3',...
    '2,1','2,2','2,3',...
    '3,1','3,2','3,3',...
```

```
         '4,1','4,2','4,3',...
         '5,1','5,2','5,3'});
xlabel('Cell Lines','fontsize',15,'fontname','herculanum')
ylabel('Conditions ({\it Temp., Conc.})','fontsize',12,...
         'fontname','courier')
zlabel('Number of Cells ','fontsize',16,'fontname','times')
view(-100,30)
set(handleAxes2,'fontsize',12)
```

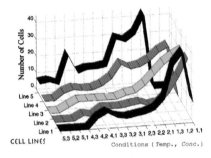
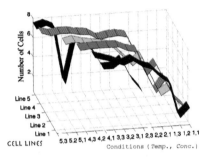

We now have the two sets of results displayed next to each other. At this stage we realise that the angle we selected may not be the best to visualise the trend of the distances as some of the ribbons overlap each other. We can easily change the angles of both graphs (or view them with different angles if necessary).

```
set(handleAxes ,'view',[105 65])
set(handleAxes2,'view',[105 65])
```

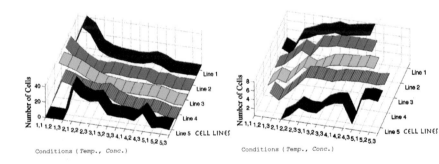

It is for this kind of modifications where it is useful to have the handles of the plots and individual elements of figures, thus we are able to change the parameters of the handles instead of having to replot things again and again.

Another very useful tool when generating figures is the use of annotations: lines, arrows, text boxes, etc. To add annotations to a figure we use the command 'annotation' with a series of parameters. First, we need to define the figure where we want the annotation by using the handles previously defined or 'gcf'. Then we specify the type of annotation out of the following options: 'line', 'arrow', 'textarrow', 'textbox', 'rectangle', 'ellipse'. We will include two arrows with text to highlight a possible outlier in the results of the experiment. The arrows are defined in positions 'X' and 'Y', again in a position relative to the whole figure. We can specify the colour and width of the line; the default values are black and 0.5 for a thin line. Finally we assign the text associated with the arrows. Since we have mentioned that Matlab can interpret LaTeX-style commands, we will show how to use a common case for symbols, namely the Greek alphabet. We can also change the colour, size and type of font. So, here are the commands for the two arrows:

```
arrow1 = annotation(handleFigure,'textarrow',...
    [0.39 0.32],[0.69 0.37],'linewidth',3,...
    'string','(\alpha)','fontsize',30);
arrow2 = annotation(handleFigure,'textarrow',...
    [0.86 0.79],[0.69 0.37],'linewidth',3,...
    'string','(\beta)','fontsize',30,'textcolor','r');
```

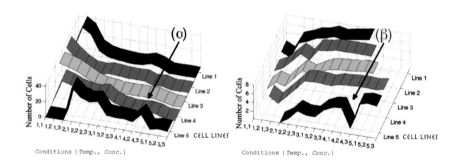

We will introduce a final refinement to the figure. In some cases, it is useful to have extra elements on the graphics, for instance to highlight a certain level above or below which something is acceptable or not. Let's suppose that for our example we are interested in a minimum number of five cells per region of interest and a minimum distance of 2.7. We would like to show a plane over each of the plots to visualise these conditions. To add a plane, we need to define the (x,y,z) coordinates; in this example, the coordinates will span from 1 to 5, 1 to 15 in the X and Y axes. We can quickly generate the coordinates with the command 'meshgrid':

```
[rows_2D,columns_2D]            = meshgrid((1:15),(0.5:5.5));
```

The corresponding coordinates for the 'z' axis will be a series of ones with same dimensions as the results from the command 'meshgrid':

```
levels_2D                       = ones(size(rows_2D));
```

We can always introduce more elements to our figures. We need to specify that we want to do this, and not to clear the current axes and start again, with the command 'hold on'. We will first introduce the plane in the left-hand side plot:

```
subplot(handleAxes)
hold on
```

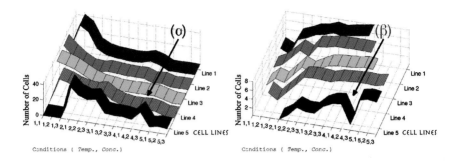

We will now display the plane as a surface with the command 'surf'. We select green colour for the face and no edges between the cells of the plane. The height of the plane is determined as one of the parameters to the command 'surf' by multiplying the variable levels_2D by the value of the number of cells, five in this case.

```
errSurface1 = surf(columns_2D,rows_2D,5*levels_2D,...
    'edgecolor','none','facecolor','g');
```

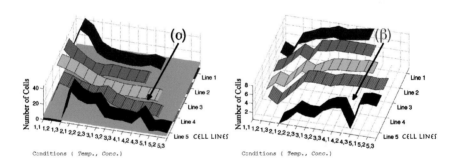

We can change the transparency of the plane so that the elements below are still visible. At the same time we will add a title to identify the left and right parts of the figure:

```
set(errSurface1,'facealpha',0.75)
title('(a)','fontsize',20)
```

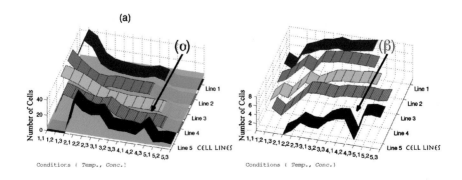

We can now add a plane on the second axis of the figure. Notice that we will modify the height of the plane to 2.7 as was specific to the distance between the cells.

```
subplot(handleAxes2)
hold on
errSurface2 = surf(columns_2D,rows_2D,2.7*levels_2D,...
    'edgecolor','none','facecolor','g');
set(errSurface2,'facealpha',0.75)
title('(b)','fontsize',20)
```

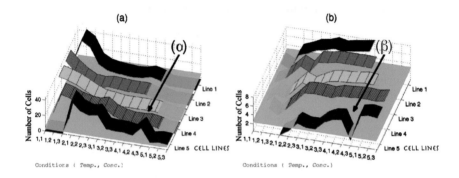

With all these commands saved as a Matlab file (Figure2_JournalB.m), it is sufficient to type the name of the script on the Command Window or to run the file from the editor to generate the whole figure. If modifications are required, they can be done on the file and then the figure is re-created in one click.

8.4   Recipe 4: Three-Dimensional Graphics

We will now explore a few techniques to use with 4D data sets, like the ones that are generated by confocal or multi-photon microscopes. These microscopes use a laser beam that scans over a plane of a sample, acquiring a 2D image one point at a time. The microscopes have the ability to concentrate the power of the laser at a certain depth; therefore, it is possible to acquire planes at different depths so as the process is repeated, a 3D volume is acquired. Furthermore, this 3D acquisition process can be repeated time after time so that a series of 3D sets correspond to different time points of a 4D data set.

In this recipe we will explore some techniques to visualising 3D sets. The next recipe will extend these techniques to the 4D case. As an example, we will load a synthetic dataset that resembles a series of vessels in which a fluorescent substance is flowing.

```
load('dataPermeability.mat')
```

In subsequent steps we will need to know the dimensions of the set to be able to plot in the correct coordinates, so the first step is to calculate the size of the set and to generate 2D and 3D coordinates. The 2D coordinates are easily generated with the command 'meshgrid', and we will generate a matrix with ones to be used for the third dimension:

```
[rows,columns,levels,timePoints]   = size(dataPermeability);
[rows_2D,columns_2D]               = meshgrid((1:columns),...
                                       (1:rows));
levels_2D                          = ones(size(rows_2D));
```

Later on we will need to take some decisions based on the values of the data set, so we will investigate its maximum and minimum values:

```
[min(dataPermeability(:)) max(dataPermeability(:))]
```

```
    ans =

      -38.3889   240.3411
```

For the 3D coordinates we can repeat the matrices for the rows and the columns, but for the levels in the 'z' axis, we need to increase the values by one per level. We do this with a loop from 1 to the number of levels previously calculated:

```
rows_3D                         = repmat(rows_2D,[1 1 20]);
columns_3D                      = repmat(columns_2D,[1 1 20]);
for k=1:levels
    levels_3D(:,:,k)            = levels_2D*k;
end
```

With the coordinates in place, we can now explore a few options to display 3D data sets. The first graph will show a series of slices one time point of the data, each on top of another, and to reveal more of the data we will add some transparency with the command 'alpha'. The image is displayed with the command 'surf', which is useful to display an image in a 3D context.

```
figure
hold on
for counterLevels =1:levels
    surf(rows_2D(1:4:end,1:4:end),...
        columns_2D(1:4:end,1:4:end),...
        counterLevels*levels_2D(1:4:end,1:4:end),...
        dataPermeability(1:4:end,1:4:end,counterLevels,1),...
        'edgecolor','none');
end
view(145,50)
axis tight
alpha(0.3)
```

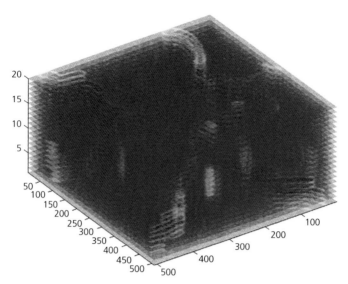

The previous graph shows roughly how the data is distributed and where some of the 'vessels' are located. However, depending on the data, this graph may not be very clear, for instance, if the top slices are empty of any detail. Another display option is to "slice" the data so that we remove sections that may be hiding interesting details. To create 3D plots with "slices", we can combine the commands 'isocaps' and 'patch' in the following way:

```
figure
permeabilityCaps   = isocaps(rows_3D(1:4:end,1:4:end,1:10), ...
    columns_3D(1:4:end,1:4:end,1:10), ...
    levels_3D(1:4:end,1:4:end,1:10), ...
    dataPermeability(1:4:end,1:4:end,1:10,1),   -40);
patch1              = patch(permeabilityCaps,...
    'FaceColor','interp','EdgeColor','none');
view(145,50)
axis tight
```

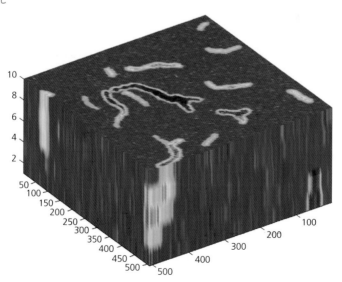

The command 'isocaps' takes as the first three input parameters the 3D coordinates corresponding to the data, as fourth, the data to be "capped" at a certain value, which is the fifth and last parameter. The minimum value of 'dataPermeability' is −38.3889. Therefore, since we wish to include all the details of the data we select a threshold slightly lower than that.

The output of 'isocaps' is passed as the first parameter to 'patch' together with some details for the plot, such as no edges to avoid a "meshy" look and an interpolation between the faces of the pixels. Notice that we have subsampled the data by a factor of 4 in the row and column directions, as these type of graphs are computationally intensive and thus we will reduce the computation time by the subsampling the data.

The first "patch" covers all the rows and columns for the first 10 levels in the 'z' axis. We will now add two blocks in the upper 10 levels. We can combine the two commands in a single line.

```
patch2     = patch(isocaps(rows_3D(1:4:end,1:4:120,10:20),...
    columns_3D(1:4:end,1:4:120,10:20),...
    levels_3D(1:4:end,1:4:120,10:20),...
    dataPermeability(1:4:end,1:4:120,10:20,1), -40),...
    'FaceColor','interp','EdgeColor','none');
patch3     = patch(isocaps(rows_3D(1:4:100,1:4:end,10:18),...
    columns_3D(1:4:100,1:4:end,10:18),...
    levels_3D(1:4:100,1:4:end,10:18),...
    dataPermeability(1:4:100,1:4:end,10:18,1), -40),...
    'FaceColor','interp','EdgeColor','none');
view(145,50)
axis tight
```

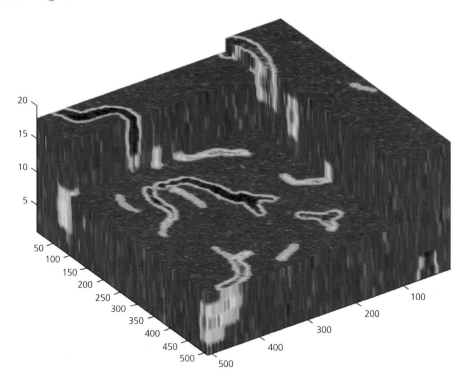

This slicing technique is quite useful in many cases, but the results are still limited by the planes that appear in the figure, and these planes have to be manually selected.

There is another way in which we can explore the 3D data without having the problem of planes blocking the view of voxels of interest that lay behind them. This new technique will remove all the voxels of a volume that do not satisfy a condition. In this case it will be an intensity threshold and will display only those voxels above the threshold.

First, we will define a suitable threshold for the data and will compare the 3D time point to determine which voxels are above the threshold. Since the maximum value of the data is 240, a suitable threshold can be 120.

```
dataThreshold         = 120;
currentTimePoint      = dataPermeability(:,:,:,1);
dataAboveThreshold    = currentTimePoint>dataThreshold;
```

Next, we will find the location of those voxels; we can do this with the command 'find', which will return the linear indices of the voxels. The indices need to be converted to the actual 3D coordinates with the command 'ind2sub', which finds subscripts (like rows and columns) from the linear indices.

```
pointsAboveThreshold  = find(dataAboveThreshold);
[c,r,l]               = ind2sub([rows,columns,levels],...
                            pointsAboveThreshold);
```

Finally, we can display the voxels using the command 'plot3', which is the 3D version of the command 'plot' with the coordinates previously located:

```
figure
plot3(r,c,l,'r.')
grid on
axis([1 rows 1 columns 1 levels])
view(110,80)
```

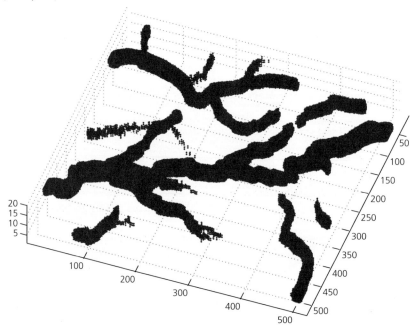

In the this graph, we only display those elements of our interest, and whilst the display is quite different from the previous ones since it is not displaying the data in planes, the voxels of interest can be quite interesting and reveal much about the data.

This technique of displaying voxels is quite powerful as we can subdivide those voxels into groups and display them with different properties. We will repeat the previous example, but in this case, we will detect regions of voxels that are contiguous to each other with the command 'bwlabeln', which is equivalent to 'bwlabel' but for more than two dimensions. The command also returns the number of labelled regions.

```
dataThreshold                        = 120;
currentTimePoint                     = ...
    dataPermeability(:,:,:,1);
dataAboveThreshold                   = ...
    currentTimePoint>dataThreshold;
[dataAboveThreshold_Label,numLab]    = ...
    bwlabeln(dataAboveThreshold);
```

Next, we will analyse the areas of the labelled regions with the command 'regionprops'. These areas will be used so that we can display only the largest regions of connected voxels. We will find those areas by sorting the regions by their "areas", which correspond to the number of voxels in each label.

```
AreaRegions                      = ...
    regionprops(dataAboveThreshold_Label,'Area');
[individualAreas,indexAreas]     = ...
    sort([AreaRegions.Area],'descend');
```

Now we can loop over the number of regions we wish to display. For this example, we select 10. Since we want the regions to be visibly distinct, we will assign them markers with different shapes (x, o, +, *, ., d-diamond, s-square) and colours (b, r, k, m, g, y, c). We will store those conditions in an array, which will later be passed as an input argument to the plot. We will loop for the number of regions that we have determined and select their 3D coordinates in the same way as before, and then display them with one of the marker-colour combination.

```
markerArray ={'bx','ro','k+','m*','g.','yd','cs'};
figure
hold on
for counter = 1:10
    pointsAboveThreshold = ...
        find(dataAboveThreshold_Label==indexAreas(counter));
    [c,r,l] = ...
        ind2sub([rows,columns,levels],pointsAboveThreshold);
    plot3(r,c,l,markerArray{1+mod(counter,7)})
end
grid on
axis([1 rows 1 columns 1 levels])
view(110,80)
```

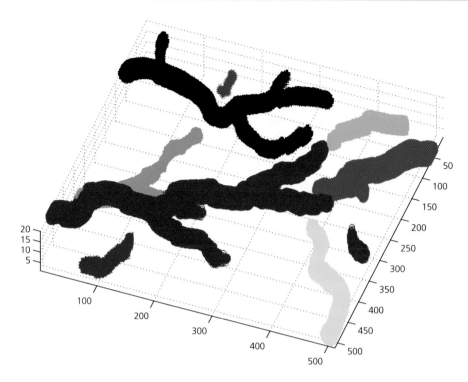

Notice that the array for the markers has only 7 elements and we displayed 10 regions. What we did was use a mathematical function that always returns a value between 1 and 7 so that no matter how many regions we plot, we always use one of the values of the array. The function 'mod' is called Modulus after division, and returns the remainder of a division, for instance:

```
mod([1 2 3 4 5 6 7 8 9],7)
```

```
     ans =

       1     2     3     4     5     6     0     1     2
```

Thus, by obtaining the mod of the counter, we will have values between 0 and 6; if we add 1, the values will be between 1 and 7. One advantage of this type of graph, by plotting points, compared with surfaces and other rendering techniques, is that it is faster and less computationally intensive.

Finally, we will create a combined graphic with a plane and a surface. First, we will use the command 'isosurface', which is similar to 'isocaps', but generates a surface above a defined threshold from a 3D data set. Again, we will combine 'isosurface' with 'patch', and we will subsample the data to reduce the computational complexity.

```
figure
p = patch(isosurface(rows_3D(1:4:end,1:4:end,:),...
    columns_3D(1:4:end,1:4:end,:),...
    levels_3D(1:4:end,1:4:end,:),...
    dataPermeability(1:4:end,1:4:end,:,1),90));
```

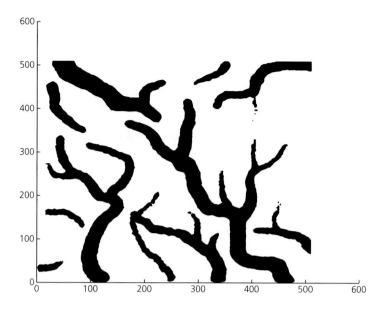

The graph can be modified by adding lights to the scene, defining a lighting mode and setting a colour for the surface. The following lines will give a different view to the surface. First we will set the face of the surfaces generated by the commands 'patch' and 'isosurface' with a red colour:

```
set(p,'FaceColor','red');
```

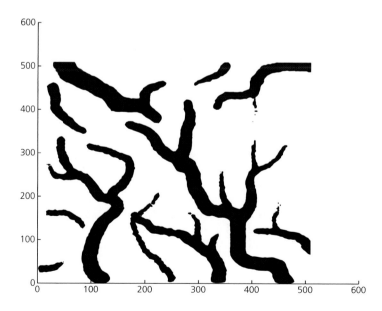

Next we will remove the colour of the edges as this will create a much better figure, which will be seen soon.

```
set(p,'FaceColor','red','EdgeColor','none');
```

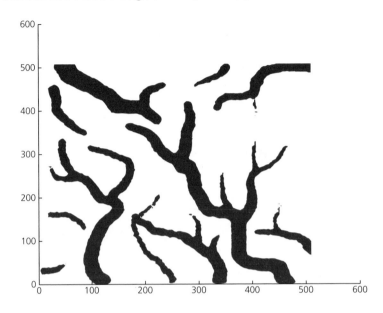

Now we add a light to the scene with the command 'camlight' and then we select the lighting algorithm with the command 'lighting' followed by one of the following options: 'none', 'flat', 'phong', 'gouraud'. In this case, we will use 'gouraud'. We also adjust the view and add a grid.

```
camlight
lighting gouraud
grid on
axis tight
view(110,80)
```

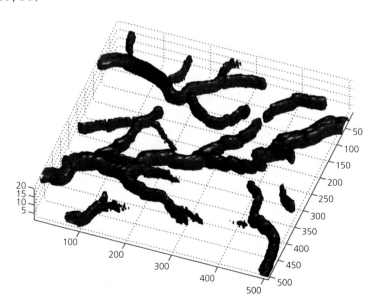

To give more context to the surface, we can add a slice of the raw intensity data in the same way that we presented the slices at the beginning of this section:

```
hold on
surf(rows_2D(1:4:end,1:4:end),...
    columns_2D(1:4:end,1:4:end),10*...
    levels_2D(1:4:end,1:4:end),...
    dataPermeability(1:4:end,1:4:end,10,1),...
    'edgecolor','none');
```

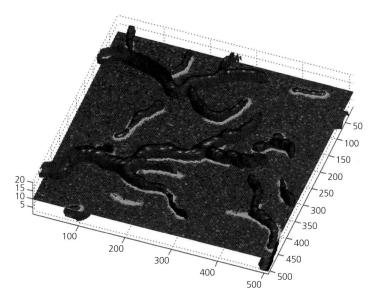

We will add transparency to the graph with the command 'alpha' so that surfaces below the plane are visible.

```
alpha(0.7)
```

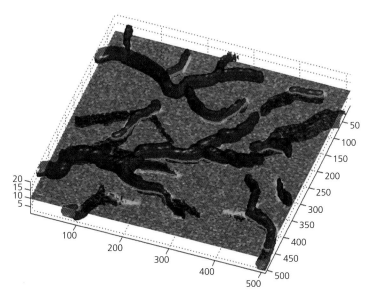

We can see that with a small number of instruction lines we can create a high quality figure. As it has been mentioned before, we can save these commands into an 'm-file', and we could repeat the figure with one click with different data.

8.5   Recipe 5: Projections

In the previous recipe we investigated several ways to display data in three dimensions. Whilst these graphics are useful, there are cases in which we would like to use 2D images, which reflect the contents of the whole three or more dimensions of the set. One way of achieving this is through "projections", which allow the analysis of the data through one particular dimension of the data.

Perhaps the most common projection is the "Maximum Intensity Projection", in which a ray traces a 3D volume and records the maximum value of the data it passes through. The ray scans the data in a raster way so that a new image is created with the maximum values of each position of the ray.

Of course, the projection techniques are not restricted to the maximum; other mathematical operations can be used. To visualise these projections we will use a 4D data set of synthetic vasculature.

```
load('dataPermeability.mat')
```

This data set has 9 time points of 3D data:

```
size(dataPermeability)
```

```
        ans =

            512    512    20    9
```

To obtain the maximum intensity projection, we first need to decide the time point, say the first, and the dimension of the projection. For instance, we want to project over the third dimension of the data, that is, the 'z' axis.

```
maxIntensityProj_Z = max(dataPermeability(:,:,:,1),[],3);
aveIntensityProj_Z = mean(dataPermeability(:,:,:,1),3);
minIntensityProj_Z = min(dataPermeability(:,:,:,1),[],3);
```

Notice that the syntax for the command 'mean' is different from that of the command 'max' and 'min'. To obtain the mean, the second parameter of the function directly indicates the dimension of the projection, whilst with 'max' and 'min' it is necessary to pass the dimension as the third parameter and pass an empty parameter [] as the second. Other functions that are useful are 'median' or 'std' to obtain the standard deviation over the projection. We can now observe the projections:

```
figure
subplot(1,3,1)
imagesc(maxIntensityProj_Z)
colorbar
subplot(1,3,2)
imagesc(aveIntensityProj_Z)
colorbar
```

```
subplot(1,3,3)
imagesc(minIntensityProj_Z)
colorbar

set(gcf,'position',[ 72    510    984    296])
```

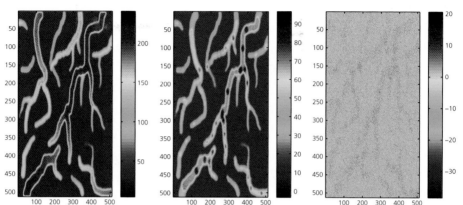

We can also obtain the projections over other dimensions, such as rows, columns or even time points:

```
maxIntensityProj_R = max(dataPermeability(:,:,:,1),[],1);
maxIntensityProj_C = max(dataPermeability(:,:,:,1),[],2);
maxIntensityProj_T = min(dataPermeability(:,:,8,:),[],4);
```

If we observe the dimensions of the first two projections, we can see that these have collapsed one of the dimensions:

```
size(maxIntensityProj_R)
size(maxIntensityProj_C)
```

```
        ans =

            1    512    20


        ans =

          512      1    20
```

If we wish to display these projections as images, we need to reshape these matrices to be two-dimensional; this can be done with the command 'squeeze'. We can thus recalculate the projections concatenating the commands:

```
maxIntensityProj_R = squeeze(max(dataPermeability(:,:,:,1),...
                       [],1));
maxIntensityProj_C = squeeze(max(dataPermeability(:,:,:,1),...
                       [],2));
```

Notice also that it is not necessary to "squeeze" the projection over time since we selected to obtain the maximum value over time for one particular level of the data.

We can combine the projections in several dimensions with the 3D data set in a similar way as in one of the previous recipes. We will display all the levels of the data set as planes and the projections as other planes at a certain distance from the data set.

As in the other cases, we will need 2D and 3D coordinates:

```
[rows,columns,levels,timePoints]    = size(dataPermeability);
[rows_2D,columns_2D]                = ...
    meshgrid((1:columns),(1:rows));
levels_2D                           = ones(size(rows_2D));

rows_3D                             = repmat(rows_2D,[1 1 20]);
columns_3D                          = ...
    repmat(columns_2D,[1 1 20]);
for k=1:levels
    levels_3D(:,:,k)                = levels_2D*k;
end
```

We can now display the data set with the command 'surf' at different levels, but first we will display three lines (or "rays") to indicate the direction of the projections.

```
figure
hold on
plot3([400   400],[400   400],[40 -17],'k-^','linewidth',2)
plot3([400   400],[650 -270],[15   15],'k-^','linewidth',2)
plot3([650 -270],[400   400],[15   15],'k-^','linewidth',2)

for counterLevels =1:levels
    surf(rows_2D(1:4:end,1:4:end),...
        columns_2D(1:4:end,1:4:end),...
        counterLevels*levels_2D(1:4:end,1:4:end),...
        dataPermeability(1:4:end,1:4:end,counterLevels,1),...
        'edgecolor','none');
end

view(145,40)
alpha(0.5)
```

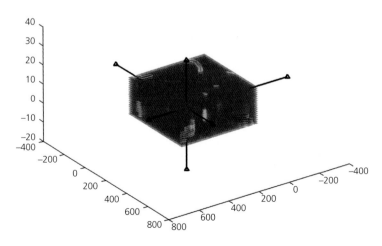

With the 3D data set in place, we can now display the projections. For the projections over the rows and columns, we need to 'squeeze' the coordinates in the same way as we did with the projections.

```
surf(rows_2D,columns_2D,-20*levels_2D(:,:),...
    maxIntensityProj_Z, ...'edgecolor','none');
surf(-280+squeeze(rows_3D(:,1,:)),...
    squeeze(columns_3D(:,1,:)),...
    squeeze(levels_3D(:,1,:)),maxIntensityProj_C,...
    'edgecolor','none');
surf(squeeze(rows_3D(1,:,:)),-280+...
    squeeze(columns_3D(1,:,:)),...
    squeeze(levels_3D(1,:,:)),maxIntensityProj_R,...
    'edgecolor','none');
axis([-280 512 -280 512 -20 20])
```

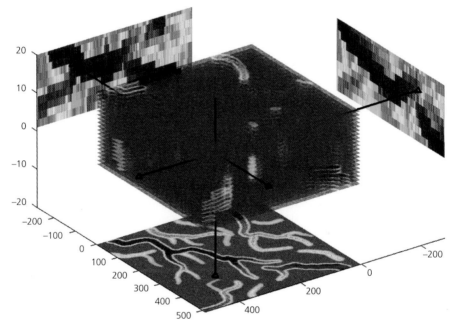

The combined graphs create an interesting figure, as some of the detail that is obscured in the 3D volume can now be observed in the projections.

8.6 Recipe 6: Four-Dimensional Data Set Explored

Creating figures with three dimensions implies the decision of what to view and how to display the data. For four dimensions, we can repeat some of the techniques previously presented in a timeline so that the variations with time become visible. We will illustrate this process with the same data sets that simulate flow of a fluorescent marker over vessels.

```
load('dataPermeability.mat')
```

The first step is to calculate the 3D coordinates that will be used for displaying the data in three dimensions, and for this we need to know the dimensions of the data set. In addition we will save the maximum and minimum values of the data sets, as these will be necessary later.

```
[rows,columns,levels,timePoints]    = size(dataPermeability);
[rows_2D,columns_2D]                = ...
    meshgrid((1:columns),(1:rows));
levels_2D                           = ones(size(rows_2D));
rows_3D                             = repmat(rows_2D,[1 1 20]);
columns_3D                          = ...
    repmat(columns_2D,[1 1 20]);
for k=1:levels
    levels_3D(:,:,k)                = levels_2D*k;
end

minValue                            = min(dataPermeability(:));
maxValue                            = max(dataPermeability(:));
```

We will create the figure and extend it over the horizontal dimension so that it can appear as a timeline. Since our data set consist of nine 3D sets, it is possible to display all of the time points. However, for a data set with more time points, say 100, it would not be possible nor desirable to display all of them; in those cases, a reduced number of time points may be selected.

The time line of our example contains nine columns, each for a time point, and three rows. In each we will display the data in different ways. The top row will display one slice of the raw intensity data. The second row will present the "maximum intensity projection", which was introduced in previous recipes. This is calculated by finding the maximum value along the 'z' dimension for each row-column position. This projection will reveal details that are normally not visible in a 3D data set. The last row will display the surfaces determined by a threshold value as we did previously. Again, we will subsample the data to reduce the computational complexity.

We need to be careful about the possible variations of the intensity from time to time, as the command 'imagesc' will scale the data from the minimum to the maximum value present at the data to be displayed. Therefore to have a consistent display along time, we will fix all the time points to have the same scale of intensities. This is achieved by using the command 'caxis' to set manually the scaling of the colour of the data. Since we have previously calculated the maximum and minimum values of the data set, we will use these values for the first and second rows.

```
figure
set(gcf,'position',[ 40   200   1300 500])

for counterPoints=1:timePoints
    subplot(3,timePoints,counterPoints)
    imagesc(dataPermeability(:,:,5,counterPoints));
    caxis([minValue maxValue])
    title(strcat('Time = ',num2str(counterPoints)),...
        'fontsize',13);
    set(gca,'xtick',[0:200:600],'ytick',[0:200:600]);
    %axis off
    subplot(3,timePoints,timePoints+counterPoints)
```

```
imagesc(max(dataPermeability(:,:,:,counterPoints),[],3));
set(gca,'xtick',[0:200:600],'ytick',[0:200:600]);
caxis([minValue maxValue])
%axis off
subplot(3,timePoints,2*timePoints+counterPoints)

p= patch(isosurface(rows_3D(1:4:end,1:4:end,:),...
    columns_3D(1:4:end,1:4:end,:),...
    levels_3D(1:4:end,1:4:end,:),...
    dataPermeability(1:4:end,1:4:end,:,...
    counterPoints),90));
% The graph can be modified by adding lights to the scene,
% defining a lighting mode and setting a colour for the
% surface. The following lines will give a different
% view to the surface:

camlight
lighting gouraud
grid on
set(gca,'xtick',[0:200:600],'ytick',[0:200:600],...
    'ztick',[0:10:20]);
axis tight
view(110,80)
set(p,'FaceColor','red','EdgeColor','none');
end
```

```
colormap(gray)
```

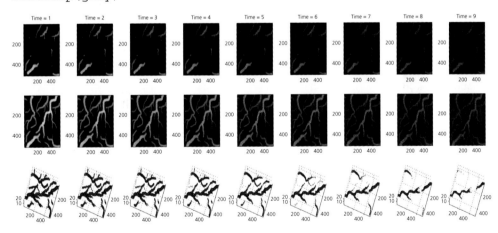

Notice that we modified the number of ticks along the axes to have a cleaner graph with fewer numbers. Also, we added the time points as the titles of the top row. We need to add labels to the left column to indicate what is being displayed in each row:

```
subplot(3,timePoints,1)
ylabel('Level = 5','fontsize',13)
```

```
subplot(3,timePoints,timePoints+  1)
ylabel('Maximum Intensity Projection','fontsize',12)
subplot(3,timePoints,2*timePoints+1)
zlabel('Surface above 90','fontsize',13)
```

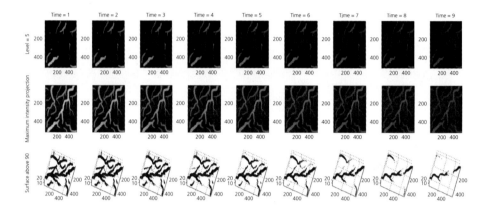

The timeline indicates how the intensity decreases with time in three different ways.

Another simple way to display the variation of the intensity with time is to select one column (say 100 or 300) of one level (say 5) for every time point and to display the columns of each time point next to each other. In this way, the vertical axis will correspond to the rows of the data, and the horizontal axis will correspond to the time points.

To help the visualisation of the variation of the intensity we can display together the intensity data from which the columns are being selected, together with the column arrangement. In the following figure we have selected two columns, 100 and 300, and changed the values on the intensity to highlight their position in the figure.

```
figure
set(gcf,'position',[ 40   300   1200 300])

subplot(131)
currentSlice                        = dataPermeability(:,:,5,1);
currentSlice(:,[96:99   296:299 ]) = maxValue;
imagesc(currentSlice)
caxis([minValue maxValue])
xlabel('Columns','fontsize',16)
ylabel('Rows','fontsize',16)
subplot(132)
imagesc(squeeze(dataPermeability(:,100,5,:)))
caxis([minValue maxValue])
xlabel('Time points','fontsize',16)
ylabel('Rows','fontsize',16)
subplot(133)
imagesc(squeeze(dataPermeability(:,300,5,:)))
```

```
caxis([minValue maxValue])
xlabel('Time points','fontsize',16)
ylabel('Rows','fontsize',16)
colorbar
```

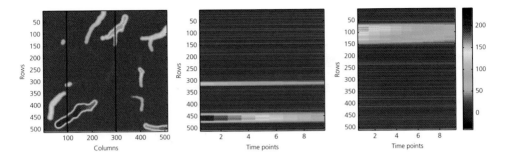

This way of displaying the time, as the horizontal axis of a 2D graph, is very useful as it shows clearly the variation of intensity, which could be determined by the concentration of a fluorophore, for instance, as time passes in an experiment.

As a final way to display the variation of the intensity with time, we can create surfaces that correspond to the same threshold level at different points in time. This is similar to what is presented in the bottom row of the previous timeline, but in a single graph. In this way, it is easier to compare the variation.

Since the surfaces from nine time points would create a very complex figure, we have selected only three time points: 1, 6 and 9. We will display them in reverse order and assign different colours and increasing levels of transparency so that it is clear which surfaces are contained inside others. As before, we subsample the data taking one out of every four elements in the rows and columns to reduce the complexity.

```
figure

p1 = patch(isosurface(rows_3D(1:4:end,1:4:end,:),...
     columns_3D(1:4:end,1:4:end,:),...
     levels_3D(1:4:end,1:4:end,:),...
     dataPermeability(1:4:end,1:4:end,:,9),80));
set(p1,'FaceColor','blue','EdgeColor','none','facealpha',0.9);

p2 = patch(isosurface(rows_3D(1:4:end,1:4:end,:),...
     columns_3D(1:4:end,1:4:end,:),...
     levels_3D(1:4:end,1:4:end,:),...
     dataPermeability(1:4:end,1:4:end,:,6),80));
set(p2,'FaceColor','red','EdgeColor','none','facealpha',0.4);

p3 = patch(isosurface(rows_3D(1:4:end,1:4:end,:),...
     columns_3D(1:4:end,1:4:end,:),...
     levels_3D(1:4:end,1:4:end,:),...
     dataPermeability(1:4:end,1:4:end,:,1),80));
set(p3,'FaceColor','green','EdgeColor','none',...
     'facealpha',0.2);
```

```
camlight
lighting gouraud
grid on
axis tight
view(110,80)
```

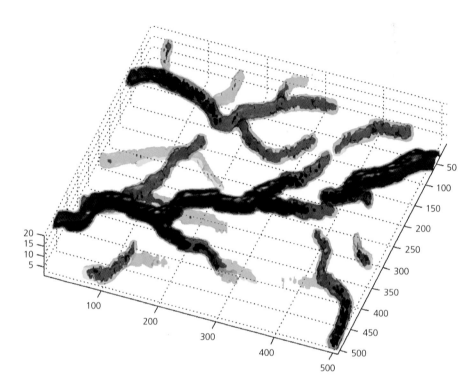

Whilst with all the previous figures it was clear that the intensity was decreasing along the time, in this last figure it becomes clear how some regions, and not others, disappear below the selected threshold (80) with time.

Another way to display the changing surfaces is to use a "mesh" above a "surface". This can be easily achieved by changing the settings of the patches. In the following figure we remove the face colour and assign a colour with transparency to the edges. The visual effect is very different from using only surfaces.

```
figure

p1 = patch(isosurface(rows_3D(1:4:end,1:4:end,:),...
     columns_3D(1:4:end,1:4:end,:),...
     levels_3D(1:4:end,1:4:end,:),...
     dataPermeability(1:4:end,1:4:end,:,9),80));
set(p1,'FaceColor','blue','EdgeColor','none','facealpha',0.9);

p3 = patch(isosurface(rows_3D(1:4:end,1:4:end,:),...
     columns_3D(1:4:end,1:4:end,:),...
```

```
        levels_3D(1:4:end,1:4:end,:),...
        dataPermeability(1:4:end,1:4:end,:,1),80));
set(p3,'FaceColor','none','EdgeColor','red','edgealpha',0.1);


camlight
lighting gouraud
grid on
set(gca,'xtick',[0:200:600],'ytick',[0:200:600],...
    'ztick',[0:10:20]);
axis tight
view(110,80)
```

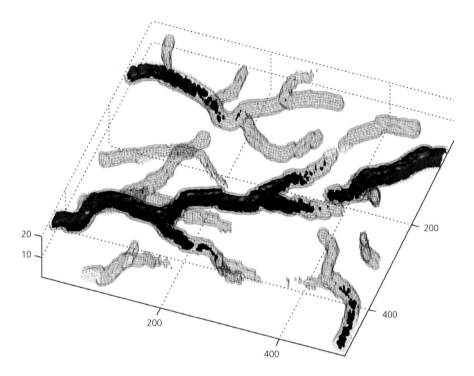

As with the previous figure, it is clear which regions have changed over time and in which regions, the concentration of the fluorophore has remained above the threshold.

A final variation of this figure is to position partial results at different heights; in this way, the variation in time is contained with the vertical axis. We can of course display not only meshes but also images within this vertical arrangement.

```
figure
hold on
p1 = patch(isosurface(rows_3D(1:4:end,1:4:end,:),...
    columns_3D(1:4:end,1:4:end,:),...
    150+levels_3D(1:4:end,1:4:end,:),...
```

```
        dataPermeability(1:4:end,1:4:end,:,9),80));
set(p1,'FaceColor','none','EdgeColor','magenta',...
    'edgealpha',0.1);


p2 = patch(isosurface(rows_3D(1:4:end,1:4:end,:),...
    columns_3D(1:4:end,1:4:end,:),...
    100+levels_3D(1:4:end,1:4:end,:),...
    dataPermeability(1:4:end,1:4:end,:,5),80));
set(p2,'FaceColor','none','EdgeColor','blue','edgealpha',0.1);


p3 = patch(isosurface(rows_3D(1:4:end,1:4:end,:),...
    columns_3D(1:4:end,1:4:end,:),...
    50+levels_3D(1:4:end,1:4:end,:),...
    dataPermeability(1:4:end,1:4:end,:,1),80));
set(p3,'FaceColor','none','EdgeColor','red','edgealpha',0.1);


surf(rows_2D(1:4:end,1:4:end),...
    columns_2D(1:4:end,1:4:end),-50+1*...
    levels_2D(1:4:end,1:4:end),...
    max(dataPermeability(1:4:end,1:4:end,:,1),[],3),...
        'edgecolor','none');


contour(rows_2D(1:4:end,1:4:end),...
    columns_2D(1:4:end,1:4:end),...
    max(dataPermeability(1:4:end,1:4:end,:,1),[],3),...
        [40 100 200 300]);


camlight
lighting gouraud
grid on
set(gca,'xtick',[0:200:600],...
    'ytick',[0:200:600],...
    'ztick',[1 65 115 165],...
    'ztick',[-49 1 65 115 165],...
    'zticklabel',['MIP  t=1';'Cont t=1';...
    't=1      ';'t=5     ';'t=9      '],...
    'fontsize',15);
set(gcf,'position',[500   200   600   800]);
axis tight
view(155,10)
rotate3d on
```

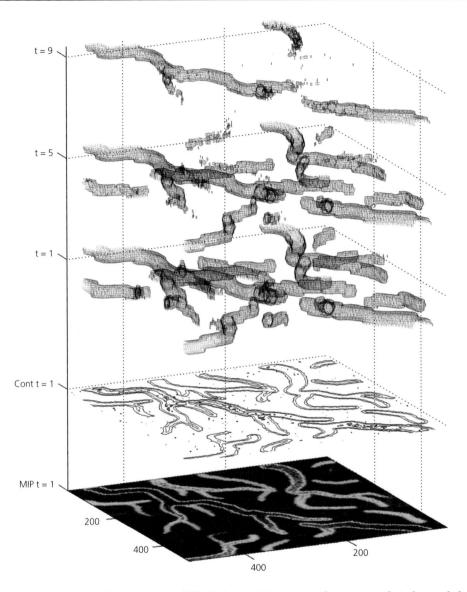

In the previous figure, we modified the position over the z-axis of each patch by adding multiples of 50 to give a good separation between three time points as each time point consisted of nine levels. We changed the edge colour for each time point and removed the face colours to have a cleaner figure. In addition we displayed the maximum intensity projection of the first time point together with the contours of this projection at four different intensity levels. Notice that the labels of the z-axis were typed as a matrix with one label at every row. As all the rows need to be of the same length, some have extra spaces.

As with our previous examples, a few lines of code can generate high quality figures, and it is not too difficult to modify the code to generate similar images.

In this case, we can save the code to the file Figure3_JournalC.m.

Index

absolute value (*abs*), 17–22, 194, 195, 345
adding lines in a figure *see line; plot*
adding/modifying labels in figures, 21, 22, 343, 350–352 *see also xlabel; ylabel; zlabel*
adding text boxes in figures *see text* (command)
alpha (Greek alphabet), 376
alpha (transparency), 76, 377, 379, 387, 395–6
areas, 113–20, 157–64, 174, 304–9, 312, 313, 332–4, 383
arguments (to/from a function), 3, 6, 18, 19, 24, 26, 29–31, 35, 43
average *see mean*
axes, 21, 28, 30, 32, 41, 51, 53, 74, 77, 85, 145, 180, 348–52
axis, 18, 19, 21, 32, 35, 36, 41, 42, 52, 53, 90, 137, 354–6
axis labels *see xlabel; ylabel; zlabel*

backgroundcolor, 361 *see also* color, setting properties
band pass filter, 207–8
bar, 24, 28, 29, 265, 266
batch processing, 70, 316–17
BODMAS/BIDMAS, 4
bold, 371
bone (colormap), 56, 83
Boolean, 13, 45
boxplot, 320–324, 357–8
branch points, 177–8
brightness, 46, 94, 215, 271, 274
bwdist, 181, 334–5
bwlabel, bwlabeln, 112, 115, 157, 166, 305
bwmorph, 175–9, 336–8
bwperim, 118, 119

Canny edge detection, 254, 260
caxis, 267–9, 392–5
centroids, 113, 116, 308–11, 331–3

change directory (*cd*), 40
characters (*char*), 3, 11, 13
chrominance, 271
circular shift (*circshift*), 257, 260
clear, 36, 42, 70, 143
clear axis *(cla),* 36
clear figure *(clf),* 36
cmy, cmyk, 71
colon operator, 6, 14, 24, 26, 218, 278, 304
colorbar, 30, 31, 55
colormap, 54–8, 81–4, 122, 158, 159
colormapeditor, 83–4
color, setting properties, 128, 130, 277, 333, 347, 360 *see also backgroundcolor; edgecolor; facecolor*
colour space, 63, 71, 271
comments, 34, 116
contour, 91–3
contrast, 54, 68, 96, 97, 219–24
convolution (*conv, conv2*), 140
cool (colormap), 56, 82
copper (colormap), 56
cosine (*cos*), 187
cross-correlation (*xcorr, xcorr2*), 254–60
cumsum, 27, 28, 266

data types, 9, 45, 46, 70, 85, 293, 294
  see also uint8; uint16; Boolean; double (data type)
diameter, equivalent diameter, 331
diff, 123, 127
dilation *see* image dilation
dir, 247, 317, 318
directories, 47 *see also* folder
displaying multiple views in one figure *see hold; plotyy; subplot*
double (data type), 3, 13, 46, 68, 85, 98, 140

eccentricity, 113, 308, 310, 331
edgealpha, 397, 398 *see also alpha* (transparency)

Biomedical Image Analysis Recipes in MATLAB®: For Life Scientists and Engineers, First Edition.
Constantino Carlos Reyes-Aldasoro.
© 2015 John Wiley & Sons, Ltd. Published 2015 by John Wiley & Sons, Ltd.

edgecolor, 231, 255, 256 *see also* color, setting properties
edge detection, 254, 255
elements in a matrix *see numel; size*
embossing filters, 210
empty matrix, 3, 16, 249
end operator, 11, 13–16
end points, 97, 178
erosion *see* image erosion
Euler number, 113, 331
exception handling, 11

facealpha, 377, 378 *see also* alpha (transparency)
*facecolo*r, 280–282, 377, 378 *see also* color, setting properties
fft, fft2, fftn, 194
fftshift, 195
figure, 28–30, 35–41, 343–5
file naming, 167, 247, 318
filling holes, 171–3
filter cubes (optical filters), 291
filtering, 139–56, 201–12
find, 114–16, 124, 126, 256, 257
folder, 39, 40, 46–8, 247
font, 22, 352, 361, 371 *see also fontname; fontsize; fontangle*
fontangle, 352
fontname, 352, 371, 375
fontsize, 21, 351, 352
for-loop, 115, 116, 128, 129
Fourier analysis, 187–213 *see also fft, fft2, fftn; fftshift; ifft, ifft2, ifftn*
fspecial, 141, 150, 204
function, 1–3, 6, 42, 43, 48

Gabor filters, 208, 209
Gaussian, 17, 26, 28, 141, 204
gca (get current axis), 41, 42, 90
gcf (get current figure), 38, 90
get, 41, 248, 344, 345, 351
gray, 54, 56, 82, 335
graythresh, 156, 302
grid, 22, 28, 45

handles to figures/axes, 90, 102, 348, 365, 369, 372, 388 *see also gca; gcf; get*
high pass filter, 143–56, 174, 184, 206–8
histogram (*hist*), 23–5, 68, 97–103, 220, 265, 266, 276, 278, 287, 288
hold, 25, 29, 75
hot (colormap), 55, 56, 82, 158
hsv (colormap), 56, 83, 158

HSV (Hue-Saturation-Value), 84, 271–4
hsv2rgb, 271, 287
hue, 46, 84, 271–89

ideal filters, 202–10
if-end, 38
ifft, ifft2, ifftn, 198
IHC/immunohistochemistry, 215, 224, 261, 262, 278
imadjust, 219–23
image dilation, 168–71, 239, 268
image erosion, 168–71, 239, 337
imagesc, 50, 51
im2bw, 303
imclose, 171–3
imdilate, 119, 168, 169, 239, 268
imerode, 168, 169, 239
imfill, 172, 173
imfilter, 142, 144, 235
imopen, 171, 172, 239
imread, 47–9
imregionalmin, 267
imrotate, 125, 126
imshow, 49–51
imtool, 96, 97, 136, 341
inner product, 9
input arguments, 3, 6, 26, 35, 43
input function, 38, 39
intensity profile, 138–51
invert intensity *see* negative image
ismember, 114, 115, 166
isocaps, 380, 381, 384
isosurface, 384, 385
italics, 352, 371

jet (colormap), 56, 58, 81, 158, 159, 329
jpg/jpeg, 37–9, 48, 69, 94, 247

kernels *see* structural elements

labelling *see bwlabel*
labels in axes *see xticklabel; yticklabel; zticklabel*
labels in figures *see xlabel; ylabel; zlabel*
Laplacian of a Gaussian filter, 150–154
LaTeX, 351, 352, 371, 376
legend, 21, 110, 325, 355
lighting, 385, 386
line, 128, 356
line specifications, 101, 109, 110
linestyle, 356, 359, 360
linewidth, 25
linspace, 6, 162, 230
load, 70, 167

logical *see Boolean*
logspace, 7
loops *see* for-loop
low pass filter, 139–43, 148–56, 174,
    201–7, 236
luminance, 271

magnitude *see* absolute value (*abs*)
major/minor axis, 113, 331
markers (for *plot* command), 19, 25, 110,
    359, 383
mathematical morphology *see*
    morphological operators
matrix concatenation, 8, 10, 109, 217,
    249, 257
matrix multiplication, 4, 8, 9, 13, 108, 195,
    201, 202
matrix orientation, 3, 7, 9
matrix transposition, 7, 9, 68, 197
maximum (*max*), 27, 29, 122, 126, 128–32,
    388, 389
mean, 26, 27, 79, 108, 122, 183, 388, 389
medial axis transform, 176
median, 27, 388
menubar, 347, 348
mesh, 30, 35, 36, 85–91, 359, 360, 396
meshgrid, 74, 376, 379
Mexican hat, 150
m-file, 2, 3, 42, 43, 69, 340, 388
minimum (*min*), 27, 122, 183, 388, 389
mkdir (make directory), 40
montage, 325, 326
morphological closing, 171–4, 239
morphological opening, 171–4, 239
morphological operators, 154, 167–79, 235,
    239–41, 268, 336
mosaics, 247–50

negative image, 107, 108
normalisation, 28, 108, 109, 194, 221, 235,
    285, 302, 341
numel, 116, 128, 129
num2str, 112, 120, 121

ones, 5, 23, 71–6, 230, 376
openvar, 5, 95
order of precedence, 4
Otsu's thresholding, 156, 163, 213, 302,
    305, 341
outer product, 9

patch, 380, 381, 384, 385, 393–9
paths, 47, 48

perimeter, 113, 114, 119, 175, 308,
    331, 334
pixel, 45–51
plot, 18–20, 24, 27–9
plot3, 192, 309, 382, 383
plotyy, 145
Poisson distribution (*poissrnd*), 24, 28
position (of figures, subplots or objects), 36,
    41, 42, 344–6
print, 38–40, 69, 343, 362
profiles *see* intensity profile
projections, 388–94
p-values of a *t*-test, 321

quadrant filters, 211, 212

rand, 23, 158, 159
randn, 17, 23, 38
regionprops, 113, 305, 308–12, 331, 332
repmat, 230–232
resize figures/axes, 344–9
resolution, 38–41, 45, 46, 69, 343
RGB (red, green, blue), 46, 63, 71, 84, 219,
    229, 271–5, 287, 347
rgb2gray, 219, 229, 264
rgb2hsv, 272, 280
ribbon, 132, 133, 365, 372–4
rotate an image *see imrotate; rotate3d*
rotate3d, 34, 87, 132

saturation, 46, 84, 218, 222, 271–89
save, 37, 39, 40, 42, 48, 69, 70, 167, 178
scaling with *imagesc,* 298–301
scatter3, 77, 275, 276, 287, 288
scatter plots, 358, 359 *see also scatter3*
screen size, 345, 346
script, 2, 3, 42, 43, 316, 343
semicolon, 5, 10, 43, 63
set (setting handle values), 41, 42, 76, 91,
    101, 102, 344–53, 385, 386
shading (command), 32, 33
shading correction, 224–46
shortcuts, 37–42
shrinking, 337, 338 *see also bwmorph*
sign, 18, 90, 188
sine (*sin*), 17, 18, 27, 188–90
size, 7–9, 16, 17, 49, 93
skeletonising, 172, 176–8, 337, 338 *see also*
    bwmorph
smoothing *see* low pass filter
sort, 27, 158, 159
spring (colormap), 56, 83

standard deviation *see std*
std, 26, 27, 116–18, 122, 130, 131, 388
stem, 25, 29
stitching, 247–61
strcat, 112, 115, 247–9, 319
strel, 171–3
string, 48, 122, 141, 247, 322, 333, 357, 361, 371
structural elements, 167–76, 240, 268
structures, 11, 113, 306, 307, 310, 318
subplot, 35, 36, 180, 181, 348, 349, 354
subsampling, 13, 232, 275, 288, 381, 384, 392, 395
sum, 9, 27–9, 43, 44, 122, 157
summer (colormap), 56
surf, 32, 36, 74–6, 88, 89, 377–9, 390, 391

TeX *see* LaTeX
text (command), 119, 333, 343, 351, 361, 370, 371, 375, 376
textcolor, 376 *see also* color, setting properties
thickness, 334, 336, 338, 354
thinning, 176, 177, 337
thresholding, 59, 60, 66, 93, 103–7, 112, 113, 156, 161, 165–7, 245, 246, 301, 302
ticks (*xtick, ytick, ztick*), 102, 110, 220, 248, 350, 369–71, 392, 393
tif/tiff, 37, 40, 48, 69, 94, 362
title, 21, 101, 112, 141, 142, 377, 378
toolbar, 97
transparency, 76, 225, 377, 379, 387, 396
   see also alpha (transparency); *facealpha; edgealpha*

try-catch, 11
t-test (*ttest, ttest2*), 321, 340

uint8 (data type), 45, 46, 65, 66, 85, 98, 108, 119, 160, 218, 221, 229, 264, 276, 285, 292–4, 297, 303, 315
uint16 (data type), 46, 293
uiopen, 47

variable, 2–11, 16–18
vessel thickness *see* thickness
view, 34–6, 87–91
Voronoi tessellation (*voronoi*), 310–316
voxel, 45, 382, 383

watershed, 313–15
who/whos, 8, 9, 70
wildcard, 318
winter (colormap), 56

xlabel, 21
xtick see ticks (*xtick, ytick, ztick*)
xticklabel, 350, 368, 370, 374

ylabel, 21
yscale, 101, 102
ytick see ticks (*xtick, ytick, ztick*)
yticklabel, 350, 369, 371, 374
yuv, 271

zeros, 23, 71–3, 180, 202, 209
zlabel, 21
ztick see ticks (*xtick, ytick, ztick*)
zticklabel, 350, 369, 398